DEFENSE, SECURITY AND STRATEGIES

U.S. NATIONAL SECURITY: A NEW STRATEGY FOR DIPLOMACY AND ENGAGEMENT

DEFENSE, SECURITY AND STRATEGIES

Additional books in this series can be found on Nova's website under the Series tab.

Additional E-books in this series can be found on Nova's website under the E-books tab.

AMERICAN POLITICAL, ECONOMIC, AND SECURITY ISSUES

Additional books in this series can be found on Nova's website under the Series tab.

Additional E-books in this series can be found on Nova's website under the E-books tab.

DEFENSE, SECURITY AND STRATEGIES

U. S. NATIONAL SECURITY: A NEW STRATEGY FOR DIPLOMACY AND ENGAGEMENT

BELINDA M. WHITE
EDITOR

Nova Science Publishers, Inc.
New York

For permission to use material from this book please contact us:
Telephone 631-231-7269; Fax 631-231-8175
Web Site: http://www.novapublishers.com

LIBRARY OF CONGRESS CATALOGING-IN-PUBLICATION DATA

U.S. national security : a new strategy for diplomacy and engagement /
editor, Belinda M. White.
p. cm.
Includes index.
ISBN 978-1-61761-436-1 (hardcover)
1. National security--United States. 2. National Security Council (U.S.)
3. Homeland Security Council (U.S.) I. White, Belinda M.
UA23.U226 2010
355'.033573--dc22

2010035911

Published by Nova Science Publishers, Inc. † New York

CONTENTS

PREFACE

This book explores America's national security strategy, which is focused on renewing American leadership so that we can more effectively advance our interests in the 21st century. This strategy recognizes the fundamental connection between our national security, our national competitiveness, resilience, and moral example. It reaffirms America's commitment to pursue our interests through an international system in which all nations have certain rights and responsibilities, which in turn, will allow America to leverage our engagement abroad on behalf of a world in which individuals enjoy more freedom and opportunity, and nations have incentives to act responsibly, while facing consequences when they do not.

Chapter 1- Time and again in our Nation's history, Americans have risen to meet — and to shape — moments of transition. This must be one of those moments. We live in a time of sweeping change. The success of free nations, open markets, and social progress in recent decades has accelerated globalization on an unprecedented scale. This has opened the doors of opportunity around the globe, extended democracy to hundreds of millions of people, and made peace possible among the major powers. Yet globalization has also intensified the dangers we face — from international terrorism and the spread of deadly technologies, to economic upheaval and a changing climate.

For nearly a decade, our Nation has been at war with a far-reaching network of violence and hatred. Even as we end one war in Iraq, our military has been called upon to renew our focus on Afghanistan as part of a commitment to disrupt, dismantle, and defeat al-Qa'ida and its affiliates. This is part of a broad, multinational effort that is right and just, and we will be unwavering in our commitment to the security of our people, allies, and partners. Moreover, as we face multiple threats — from nations, nonstate actors, and failed states — we will maintain the military superiority that has secured our country, and underpinned global security, for decades.

Chapter 2- The National Security Council (NSC) was established by statute in 1947 to create an inter-departmental body to advise the President with respect to the integration of domestic, foreign, and military policies relating to the national security so as to enable the military services and the other departments and agencies of the Government to cooperate more effectively in matters involving the national security. Currently, statutory members of the Council are the President, Vice President, the Secretary of State, and the Secretary of Defense; but, at the President's request, other senior officials participate in NSC deliberations. The Chairman of the Joint Chiefs of Staff and the Director of National Intelligence are statutory advisers. In 2007 the Secretary of Energy was added to the NSC membership.

Chapter 3- In the immediate aftermath of the terrorist attacks of September 11, 2001, President George W. Bush established the Office of Homeland Security and the Homeland Security Council (HSC). In his June 2002 proposal for a Department of Homeland Security, President Bush appeared to anticipate the continued operation of both of these entities. However, the Homeland Security Act of 2002, which mandated the new department, statutorily rechartered the HSC as an agency within the Executive Office of the President (EOP). Thereafter, the HSC disappeared from the public record, and its status today remains uncertain. Recently, some have called for the merger of the HSC with the National Security Council (NSC).

In: U.S. National Security: A New Strategy for... ISBN: 978-1-61761-436-1
Editors: Belinda M. White

Chapter 1

NATIONAL SECURITY STRATEGY - MAY 2010*

Office of the President of the United States

Time and again in our Nation's history, Americans have risen to meet — and to shape — moments of transition. This must be one of those moments. We live in a time of sweeping change. The success of free nations, open markets, and social progress in recent decades has accelerated globalization on an unprecedented scale. This has opened the doors of opportunity around the globe, extended democracy to hundreds of millions of people, and made peace possible among the major powers. Yet globalization has also intensified the dangers we face — from international terrorism and the spread of deadly technologies, to economic upheaval and a changing climate.

For nearly a decade, our Nation has been at war with a far-reaching network of violence and hatred. Even as we end one war in Iraq, our military has been called upon to renew our focus on Afghanistan as part of a commitment to disrupt, dismantle, and defeat al-Qa'ida and its affiliates. This is part of a broad, multinational effort that is right and just, and we will be unwavering in our commitment to the security of our people, allies, and partners. Moreover, as we face multiple threats — from nations, nonstate

---

* This is an edited, reformatted and augmented edition of an Office of the President of the United States publication, dated May 2010.

actors, and failed states — we will maintain the military superiority that has secured our country, and underpinned global security, for decades.

Yet as we fight the wars in front of us, we must see the horizon beyond them — a world in which America is stronger, more secure, and is able to overcome our challenges while appealing to the aspirations of people around the world. To get there, we must pursue a strategy of national renewal and global leadership — a strategy that rebuilds the foundation of American strength and influence.

Our strategy starts by recognizing that our strength and influence abroad begins with the steps we take at home. We must grow our economy and reduce our deficit. We must educate our children to compete in an age where knowledge is capital, and the marketplace is global. We must develop the clean energy that can power new industry, unbind us from foreign oil, and preserve our planet. We must pursue science and research that enables discovery, and unlocks wonders as unforeseen to us today as the surface of the moon and the microchip were a century ago. Simply put, we must see American innovation as a foundation of American power.

We must also build and integrate the capabilities that can advance our interests, and the interests we share with other countries and peoples. Our Armed Forces will always be a cornerstone of our security, but they must be complemented. Our security also depends upon diplomats who can act in every corner of the world, from grand capitals to dangerous outposts; development experts who can strengthen governance and support human dignity; and intelligence and law enforcement that can unravel plots, strengthen justice systems, and work seamlessly with other countries.

The burdens of a young century cannot fall on American shoulders alone — indeed, our adversaries would like to see America sap our strength by overextending our power. In the past, we have had the foresight to act judiciously and to avoid acting alone. We were part of the most powerful wartime coalition in human history through World War II, and stitched together a community of free nations and institutions to endure a Cold War. We are clear-eyed about the challenge of mobilizing collective action, and the shortfalls of our international system. But America has not succeeded by stepping outside the currents of international cooperation. We have succeeded by steering those currents in the direction of liberty and justice — so that nations thrive by meeting their responsibilities and face consequences when they don't.

To do so, we will be steadfast in strengthening those old alliances that have served us so well, while modernizing them to meet the challenges of a new century. As influence extends to more countries and capitals, we will build new and deeper partnerships in every region, and strengthen international standards and institutions. This engagement is no end in itself The international order we seek is one that can resolve the challenges of our times — countering violent extremism and insurgency; stopping the spread of nuclear weapons and securing nuclear materials; combating a changing climate and sustaining global growth; helping countries feed themselves and care for their sick; resolving and preventing conflict, while also healing its wounds.

In all that we do, we will advocate for and advance the basic rights upon which our Nation was founded, and which peoples of every race and region have made their own. We promote these values by living them, including our commitment to the rule of law. We will strengthen international norms that protect these rights, and create space and support for those who resist repression. Our commitment to human dignity includes support for development, which is why we will fight poverty and corruption. And we reject the notion that lasting security and prosperity can be found by turning away from universal rights — democracy does not merely represent our better angels, it stands in opposition to aggression and injustice, and our support for universal rights is both fundamental to American leadership and a source of our strength in the world.

As a Nation made up of people from every race, region, faith, and culture, America will persist in promoting peace among different peoples and believes that democracy and individual empowerment need not come at the expense of cherished identities. Indeed, no nation should be better positioned to lead in an era of globalization than America — the Nation that helped bring globalization about, whose institutions are designed to prepare individuals to succeed in a competitive world, and whose people trace their roots to every country on the face of the Earth.

As a citizen, Senator, and President, I have always believed that America's greatest asset is its people — from the awe I felt as a child watching a space capsule pulled out of the Pacific, to the strength I drew from workers rebuilding their lives in Illinois, to the respect that I have for the generation of Americans who serve our country today. That is why I also believe that we must foster even deeper connections among Americans and peoples around the globe. Our long-term security will come not from our ability to instill fear in other peoples, but through our capacity to speak to their hopes. And that work will best be done through the power of the decency and dignity of the

American people — our troops and diplomats, but also our private sector, nongovernmental organizations, and citizens. All of us have a role to play.

From the birth of our liberty, America has had a faith in the future — a belief that where we're going is better than where we've been, even when the path ahead is uncertain. To fulfill that promise, generations of Americans have built upon the foundation of our forefathers — finding opportunity, fighting injustice, and forging a more perfect Union. We have also created webs of commerce, supported an international architecture of laws and institutions, and spilled American blood in foreign lands — not to build an empire, but to shape a world in which more individuals and nations could determine their own destiny, and live with the peace and dignity that they deserve.

In 2010, America is hardened by wars, and inspired by the servicemen and women who fight them. We are disciplined by a devastating economic crisis, and determined to see that its legacy is a new foundation for prosperity; and we are bound by a creed that has guided us at home, and served as a beacon to the world. America's greatness is not assured — each generation's place in history is a question unanswered. But even as we are tested by new challenges, the question of our future is not one that will be answered for us, it is one that will be answered by us. And in a young century whose trajectory is uncertain, America is ready to lead once more.

I. Overview of National Security Strategy

At the dawn of the 21st century, the United States of America faces a broad and complex array of challenges to our national security. Just as America helped to determine the course of the 20th century, we must now build the sources of American strength and influence, and shape an international order capable of overcoming the challenges of the 21st century.

The World as It Is, A Strategy for the World We Seek

To succeed, we must face the world as it is. The two decades since the end of the Cold War have been marked by both the promise and perils of change. The circle of peaceful democracies has expanded; the specter of nuclear war has lifted; major powers are at peace; the global economy has grown; commerce has stitched the fate of nations together; and more individuals can determine their own destiny. Yet these advances have been accompanied by persistent problems. Wars over ideology have given way to wars over religious, ethnic, and tribal identity; nuclear dangers have proliferated; inequality and economic instability have intensified; damage to our environment, food insecurity, and dangers to public health are increasingly shared; and the same tools that empower individuals to build enable them to destroy.

The dark side of this globalized world came to the forefront for the American people on September 11, 2001. The immediate threat demonstrated by the deadliest attacks ever launched upon American soil demanded strong and durable approaches to defend our homeland. In the years since, we have launched a war against al-Qa'ida and its affiliates, decided to fight a war in Iraq, and confronted a sweeping economic crisis. More broadly, though, we have wrestled with how to advance American interests in a world that has changed—a world in which the international architecture of the 20th century is buckling under the weight of new threats, the global economy has accelerated the competition facing our people and businesses, and the universal aspiration for freedom and dignity contends with new obstacles.

Our country possesses the attributes that have supported our leadership for decades—sturdy alliances, an unmatched military, the world's largest economy, a strong and evolving democracy, and a dynamic citizenry. Going forward, there should be no doubt: the United States of America will continue to underwrite global security—through our commitments to allies, partners, and institutions; our focus on defeating al-Qa'ida and its affiliates in Afghanistan, Pakistan, and around the globe; and our determination to deter aggression and prevent the proliferation of the world's most dangerous weapons. As we do, we must recognize that no one nation—no matter how powerful—can meet global challenges alone. As we did after World War II, America must prepare for the future, while forging cooperative approaches among nations that can yield results.

Our national security strategy is, therefore, focused on renewing American leadership so that we can more effectively advance our interests in the 21st century. We will do so by building upon the sources of our strength at home, while shaping an international order that can meet the challenges of our time. This strategy recognizes the fundamental connection between our national security, our national competitiveness, resilience, and moral example. And it reaffirms America's commitment to pursue our interests through an international system in which all nations have certain rights and respo-nsibilities.

This will allow America to leverage our engagement abroad on behalf of a world in which individuals enjoy more freedom and opportunity, and nations have incentives to act responsibly, while facing consequences when they do not.

Renewing American Leadership—Building at Home, Shaping Abroad

Our approach begins with a commitment to build a stronger foundation for American leadership, because what takes place within our borders will determine our strength and influence beyond them. This truth is only heightened in a world of greater interconnection—a world in which our prosperity is inextricably linked to global prosperity, our security can be directly challenged by developments across an ocean, and our actions are scrutinized as never before.

At the center of our efforts is a commitment to renew our economy, which serves as the wellspring of American power. The American people are now emerging from the most devastating recession that we have faced since the Great Depression. As we continue to act to ensure that our recovery is broad and sustained, we are also laying the foundation for the long term growth of our economy and competitiveness of our citizens. The investments that we have made in recovery are a part of a broader effort that will contribute to our strength: by providing a quality education for our children; enhancing science and innovation; transforming our energy economy to power new jobs and industries; lowering the cost of health care for our people and businesses; and reducing the Federal deficit.

Each of these steps will sustain America's ability to lead in a world where economic power and individual opportunity are more diffuse. These efforts are also tied to our commitment to secure a more resilient nation. Our recovery includes rebuilding an infrastructure that will be more secure and reliable in the face of terrorist threats and natural disasters. Our focus on education and science can ensure that the breakthroughs of tomorrow take place in the United

States. Our development of new sources of energy will reduce our dependence on foreign oil. Our commitment to deficit reduction will discipline us to make hard choices, and to avoid overreach. These steps complement our efforts to integrate homeland security with national security; including seamless coordination among Federal, state, and local governments to prevent, protect against, and respond to threats and natural disasters.

Finally, the work to build a stronger foundation for our leadership within our borders recognizes that the most effective way for the United States of America to promote our values is to live them. America's commitment to democracy, human rights, and the rule of law are essential sources of our strength and influence in the world. They too must be cultivated by our rejection of actions like torture that are not in line with our values, by our commitment to pursue justice consistent with our Constitution, and by our steady determination to extend the promise of America to all of our citizens. America has always been a beacon to the peoples of the world when we ensure that the light of America's example burns bright.

Building this stronger foundation will support America's efforts to shape an international system that can meet the challenges of our time. In the aftermath of World War II, it was the United States that helped take the lead in constructing a new international architecture to keep the peace and advance prosperity—from NATO and the United Nations, to treaties that govern the laws and weapons of war; from the World Bank and International Monetary Fund, to an expanding web of trade agreements. This architecture, despite its flaws, averted world war, enabled economic growth, and advanced human rights, while facilitating effective burden sharing among the United States, our allies, and partners.

Today, we need to be clear-eyed about the strengths and shortcomings of international institutions that were developed to deal with the challenges of an earlier time and the shortage of political will that has at times stymied the enforcement of international norms. Yet it would be destructive to both American national security and global security if the United States used the emergence of new challenges and the shortcomings of the international system as a reason to walk away from it. Instead, we must focus American engagement on strengthening international institutions and galvanizing the collective action that can serve common interests such as combating violent extremism; stopping the spread of nuclear weapons and securing nuclear materials; achieving balanced and sustainable economic growth; and forging cooperative solutions to the threat of climate change, armed conflict, and pandemic disease.

The starting point for that collective action will be our engagement with other countries. The cornerstone of this engagement is the relationship between the United States and our close friends and allies in Europe, Asia, the Americas, and the Middle East—ties which are rooted in shared interests and shared values, and which serve our mutual security and the broader security and prosperity of the world. We are working to build deeper and more effective partnerships with other key centers of influence—including China, India, and Russia, as well as increasingly influential nations such as Brazil, South Africa, and Indonesia—so that we can cooperate on issues of bilateral and global concern, with the recognition that power, in an interconnected world, is no longer a zero sum game. We are expanding our outreach to emerging nations, particularly those that can be models of regional success and stability, from the Americas to Africa to Southeast Asia. And we will pursue engagement with hostile nations to test their intentions, give their governments the opportunity to change course, reach out to their people, and mobilize international coalitions.

This engagement will underpin our commitment to an international order based upon rights and responsibilities. International institutions must more effectively represent the world of the 21st century, with a broader voice—and greater responsibilities—for emerging powers, and they must be modernized to more effectively generate results on issues of global interest. Constructive national steps on issues ranging from nuclear security to climate change must be incentivized, so nations that choose to do their part see the benefits of responsible action. Rules of the road must be followed, and there must be consequences for those nations that break the rules—whether they are nonproliferation obligations, trade agreements, or human rights commitments.

This modernization of institutions, strengthening of international norms, and enforcement of international law is not a task for the United States alone—but together with like-minded nations, it is a task we can lead. A key source of American leadership throughout our history has been enlightened self-interest. We want a better future for our children and grandchildren, and we believe that their lives will be better if other peoples' children and grandchildren can live in freedom and prosperity. The belief that our own interests are bound to the interests of those beyond our borders will continue to guide our engagement with nations and peoples.

Advancing Top National Security Priorities

Just as our national security strategy is focused on renewing our leadership for the long term, it is also facilitating immediate action on top priorities. This Administration has no greater responsibility than the safety and security of the American people. And there is no greater threat to the American people than weapons of mass destruction, particularly the danger posed by the pursuit of nuclear weapons by violent extremists and their proliferation to additional states.

That is why we are pursuing a comprehensive nonproliferation and nuclear security agenda, grounded in the rights and responsibilities of nations. We are reducing our nuclear arsenal and reliance on nuclear weapons, while ensuring the reliability and effectiveness of our deterrent. We are strengthening the Nuclear Non-Proliferation Treaty (NPT) as the foundation of nonproliferation, while working through the NPT to hold nations like Iran and North Korea accountable for their failure to meet international obligations. We are leading a global effort to secure all vulnerable nuclear materials from terrorists. And we are pursuing new strategies to protect against biological attacks and challenges to the cyber networks that we depend upon.

As we secure the world's most dangerous weapons, we are fighting a war against a far-reaching network of hatred and violence. We will disrupt, dismantle, and defeat al-Qa'ida and its affiliates through a comprehensive strategy that denies them safe haven, strengthens front-line partners, secures our homeland, pursues justice through durable legal approaches, and counters a bankrupt agenda of extremism and murder with an agenda of hope and opportunity. The frontline of this fight is Afghanistan and Pakistan, where we are applying relentless pressure on al-Qa'ida, breaking the Taliban's momentum, and strengthening the security and capacity of our partners. In this effort, our troops are again demonstrating their extraordinary service, making great sacrifices in a time of danger, and they have our full support.

In Iraq, we are transitioning to full Iraqi sovereignty and responsibility—a process that includes the removal of our troops, the strengthening of our civilian capacity, and a long-term partnership to the Iraqi Government and people. We will be unwavering in our pursuit of a comprehensive peace between Israel and its neighbors, including a two-state solution that ensures Israel's security, while fulfilling the Palestinian peoples' legitimate aspirations for a viable state of their own. And our broader engagement with Muslim communities around the world will spur progress on critical political and security matters, while advancing partnerships on a broad range of issues based upon mutual interests and mutual respect.

As we rebuild the economic strength upon which our leadership depends, we are working to advance the balanced and sustainable growth upon which global prosperity and stability depends. This includes steps at home and abroad to prevent another crisis. We have shifted focus to the G-20 as the premier forum for international economic cooperation, and are working to rebalance global demand so that America saves more and exports more, while emerging economies generate more demand. And we will pursue bilateral and multilateral trade agreements that advance our shared prosperity, while accelerating investments in development that can narrow inequality, expand markets, and support individual opportunity and state capacity abroad.

These efforts to advance security and prosperity are enhanced by our support for certain values that are universal. Nations that respect human rights and democratic values are more successful and stronger partners, and individuals who enjoy such respect are more able to achieve their full potential. The United States rejects the false choice between the narrow pursuit of our interests and an endless campaign to impose our values. Instead, we see it as fundamental to our own interests to support a just peace around the world—one in which individuals, and not just nations, are granted the fundamental rights that they deserve.

In keeping with the focus on the foundation of our strength and influence, we are promoting universal values abroad by living them at home, and will not seek to impose these values through force. Instead, we are working to strengthen international norms on behalf of human rights, while welcoming all peaceful democratic movements. We are supporting the development of institutions within fragile democracies, integrating human rights as a part of our dialogue with repressive governments, and supporting the spread of technologies that facilitate the freedom to access information. And we recognize economic opportunity as a human right, and are promoting the dignity of all men and women through our support for global health, food security, and cooperatives responses to humanitarian crises.

Finally, our efforts to shape an international order that promotes a just peace must facilitate cooperation capable of addressing the problems of our time. This international order will support our interests, but it is also an end that we seek in its own right. New challenges hold out the prospect of opportunity, but only if the international community breaks down the old habits of suspicion to build upon common interests. A global effort to combat climate change must draw upon national actions to reduce emissions and a commitment to mitigate their impact. Efforts to prevent conflicts and keep the peace in their aftermath can stop insecurity from spreading. Global cooperation to prevent the spread of pandemic disease can promote public health.

Implementing this agenda will not be easy. To succeed, we must balance and integrate all elements of American power and update our national security capacity for the 21st century. We must maintain our military's conventional superiority, while enhancing its capacity to defeat asymmetric threats. Our diplomacy and development capabilities must be modernized, and our civilian expeditionary capacity strengthened, to support the full breadth of our priorities. Our intelligence and homeland security efforts must be integrated with our national security policies, and those of our allies and partners. And our ability to synchronize our actions while communicating effectively with foreign publics must be enhanced to sustain global support.

However, America's greatest asset remains our people. In an era that will be shaped by the ability to seize the opportunities of a world that has grown more interconnected, it is the American people who will make the difference—the troops and civilians serving within our government; businesses, foundations, and educational institutions that operate around the globe; and citizens who possess the dynamism, drive, and diversity to thrive in a world that has grown smaller. Because for all of its dangers, globalization is in part a product of American leadership and the ingenuity of the American people. We are uniquely suited to seize its promise.

Our story is not without imperfections. Yet at each juncture that history has called upon us to rise to the occasion, we have advanced our own security, while contributing to the cause of human progress. To continue to do so, our national security strategy must be informed by our people, enhanced by the contributions of the Congress, and strengthened by the unity of the American people. If we draw on that spirit anew, we can build a world of greater peace, prosperity, and human dignity.

II. Strategic Approach

> "More than at any point in human history—the interests of nations and peoples are shared. The religious convictions that we hold in our hearts can forge new bonds among people, or tear us apart. The technology we harness can light the path to peace, or forever darken it. The energy we use can

sustain our planet, or destroy it. What happens to the hope of a single child—anywhere—can enrich our world, or impoverish it."

—President Barack Obama, United Nations General Assembly, September 22, 2009

The United States must renew its leadership in the world by building and cultivating the sources of our strength and influence. Our national security depends upon America's ability to leverage our unique national attributes, just as global security depends upon strong and responsible American leadership. That includes our military might, economic competitiveness, moral leadership, global engagement, and efforts to shape an international system that serves the mutual interests of nations and peoples. For the world has changed at an extraordinary pace, and the United States must adapt to advance our interests and sustain our leadership.

American interests are enduring. They are:

- The security of the United States, its citizens, and U.S. allies and partners;
- A strong, innovative, and growing U.S. economy in an open international economic system that promotes opportunity and prosperity;
- Respect for universal values at home and around the world; and
- An international order advanced by U.S. leadership that promotes peace, security, and opportunity through stronger cooperation to meet global challenges.

Currently, the United States is focused on implementing a responsible transition as we end the war in Iraq, succeeding in Afghanistan, and defeating al-Qa'ida and its terrorist affiliates, while moving our economy from catastrophic recession to lasting recovery. As we confront these crises, our national strategy must take a longer view. We must build a stronger foundation for American leadership and work to better shape the outcomes that are most fundamental to our people in the 21st century.

The Strategic Environment—The World as It Is

In the two decades since the end of the Cold War, the free flow of information, people, goods and services has accelerated at an unprecedented rate. This interconnection has empowered individuals for good and ill, and challenged state based international institutions that were largely designed in the wake of World War II by policymakers who had different challenges in mind. Nonstate actors can have a dramatic influence on the world around them. Economic growth has alleviated poverty and led to new centers of influence. More nations are asserting themselves regionally and globally. The lives of our citizens —their safety and prosperity—are more bound than ever to events beyond our borders.

Within this environment, the attacks of September 11, 2001, were a transformative event for the United States, demonstrating just how much trends far beyond our shores could directly endanger the personal safety of the American people. The attacks put into sharp focus America's position as the sole global superpower, the dangers of violent extremism, and the simmering conflicts that followed the peaceful conclusion of the Cold War. And they drew a swift and forceful response from the United States and our allies and partners in Afghanistan. This response was followed by our decision to go to war in Iraq, and the ensuing years have seen America's forces, resources, and national security strategy focused on these conflicts.

The United States is now fighting two wars with many thousands of our men and women deployed in harm's way, and hundreds of billions of dollars dedicated to funding these conflicts. In Iraq, we are supporting a transition of responsibility to the sovereign Iraqi Government. We are supporting the security and prosperity of our partners in Afghanistan and Pakistan as part of a broader campaign to disrupt, dismantle, and defeat al-Qa'ida and its violent extremist affiliates.

Yet these wars—and our global efforts to successfully counter violent extremism—are only one element of our strategic environment and cannot define America's engagement with the world. Terrorism is one of many threats that are more consequential in a global age. The gravest danger to the American people and global security continues to come from weapons of mass destruction, particularly nuclear weapons. The space and cyberspace capabilities that power our daily lives and military operations are vulnerable to disruption and attack. Dependence upon fossil fuels constrains our options and pollutes our environment. Climate change and pandemic disease threaten the security of regions and the health and safety of the American people. Failing states breed conflict and endanger regional and global security. Global criminal networks foment insecurity abroad and bring people and goods across our own borders that threaten our people.

The global economy is being reshaped by innovation, emerging economies, transition to low-carbon energy, and recovery from a catastrophic recession. The convergence of wealth and living standards among developed and emerging economies holds out the promise of more balanced global growth, but dramatic inequality persists within and among nations. Profound cultural and demographic tensions, rising demand for resources, and rapid urbanization could reshape single countries and entire regions. As the world grows more interconnected, more individuals are gaining awareness of their universal rights and have the capacity to pursue them. Democracies that respect the rights of their people remain successful states and America's most steadfast allies. Yet the advance of democracy and human rights has stalled in many parts of the world.

More actors exert power and influence. Europe is now more united, free, and at peace than ever before. The European Union has deepened its integration. Russia has reemerged in the international arena as a strong voice. China and India—the world's two most populous nations—are becoming more engaged globally. From Latin America to Africa to the Pacific, new and emerging powers hold out opportunities for partnership, even as a handful of states endanger regional and global security by flouting international norms. International institutions play a critical role in facilitating cooperation, but at times cannot effectively address new threats or seize new opportunities. Meanwhile, individuals, corporations, and civil society play an increasingly important role in shaping events around the world.

The United States retains the strengths that have enabled our leadership for many decades. Our society is exceptional in its openness, vast diversity, resilience, and engaged citizenry. Our private sector and civil society exhibit enormous ingenuity and innovation, and our workers are capable and dedicated. We have the world's largest economy and most powerful military, strong alliances and a vibrant cultural appeal, and a history of leadership in economic and social development. We continue to be a destination that is sought out by immigrants from around the world, who enrich our society. We have a transparent, accountable democracy and a dynamic and productive populace with deep connections to peoples around the world. And we continue to embrace a set of values that have enabled liberty and opportunity at home and abroad.

Now, the very fluidity within the international system that breeds new challenges must be approached as an opportunity to forge new international cooperation. We must rebalance our long-term priorities so that we successfully move beyond today's wars, and focus our attention and resources on a broader set of countries and challenges. We must seize on the opportunities afforded by the world's interconnection, while responding effectively and comprehensively to its dangers. And we must take advantage of the unparalleled connections that America's Government, private sector, and citizens have around the globe.

The Strategic Approach—The World We Seek

In the past, the United States has thrived when both our nation and our national security policy have adapted to shape change instead of being shaped by it. For instance, as the industrial revolution took hold, America transformed our economy and our role in the world. When the world was confronted by fascism, America prepared itself to win a war and to shape the peace that followed. When the United States encountered an ideological, economic, and military threat from communism, we shaped our practices and institutions at home—and policies abroad—to meet this challenge. Now, we must once again position the United States to champion mutual interests among nations and peoples.

Building Our Foundation

Our national security begins at home. What takes place within our borders has always been the source of our strength, and this is even truer in an age of interconnection.

First and foremost, we must renew the foundation of America's strength. In the long run, the welfare of the American people will determine America's strength in the world, particularly at a time when our own economy is inextricably linked to the global economy. Our prosperity serves as a wellspring for our power. It pays for our military, underwrites our diplomacy and development efforts, and serves as a leading source of our influence in the world. Moreover, our trade and investment supports millions of American jobs, forges links among countries, spurs global development, and contributes to a stable and peaceful political and economic environment.

Yet even as we have maintained our military advantage, our competitiveness has been set back in recent years. We are recovering from underinvestment in the areas that are central to America's strength. We have not adequately advanced priorities like education, energy, science and technology, and health care—all of which are essential to U.S.

competitiveness, long-term prosperity, and strength. Years of rising fiscal and trade deficits will also necessitate hard choices in the years ahead.

That is why we are rebuilding our economy so that it will serve as an engine of opportunity for the American people, and a source of American influence abroad. The United States must ensure that we have the world's best-educated workforce, a private sector that fosters innovation, and citizens and businesses that can access affordable health care to compete in a globalized economy. We must transform the way that we use energy—diversifying supplies, investing in innovation, and deploying clean energy technologies. By doing so, we will enhance energy security, create jobs, and fight climate change.

Rebuilding our economy must include putting ourselves on a fiscally sustainable path. As such, implementing our national security strategy will require a disciplined approach to setting priorities and making tradeoffs among competing programs and activities. Taken together, these efforts will position our nation for success in the global marketplace, while also supporting our national security capacity—the strength of our military, intelligence, diplomacy and development, and the security and resilience of our homeland.

We are now moving beyond traditional distinctions between homeland and national security. National security draws on the strength and resilience of our citizens, communities, and economy. This includes a determination to prevent terrorist attacks against the American people by fully coordinating the actions that we take abroad with the actions and precautions that we take at home. It must also include a commitment to building a more secure and resilient nation, while maintaining open flows of goods and people. We will continue to develop the capacity to address the threats and hazards that confront us, while redeveloping our infrastructure to secure our people and work cooperatively with other nations.

America's example is also a critical component of our foundation. The human rights which America has stood for since our founding have enabled our leadership, provided a source of inspiration for peoples around the world, and drawn a clear contrast between the United States and our democratic allies, and those nations and individuals that deny or suppress human rights. Our efforts to live our own values, and uphold the principles of democracy in our own society, underpin our support for the aspirations of the oppressed abroad, who know they can turn to America for leadership based on justice and hope.

Our moral leadership is grounded principally in the power of our example—not through an effort to impose our system on other peoples. Yet over the years, some methods employed in pursuit of our security have compromised our fidelity to the values that we promote, and our leadership on their behalf. This undercuts our ability to support democratic movements abroad, challenge nations that violate international human rights norms, and apply our broader leadership for good in the world. That is why we will lead on behalf of our values by living them. Our struggle to stay true to our values and Constitution has always been a lodestar, both to the American people and to those who share our aspiration for human dignity.

Our values have allowed us to draw the best and brightest to our shores, to inspire those who share our cause abroad, and to give us the credibility to stand up to tyranny. America must demonstrate through words and deeds the resilience of our values and Constitution. For if we compromise our values in pursuit of security, we will undermine both; if we fortify them, we will sustain a key source of our strength and leadership in the world—one that sets us apart from our enemies and our potential competitors.

Pursuing Comprehensive Engagement

Our foundation will support our efforts to engage nations, institutions, and peoples around the world on the basis of mutual interests and mutual respect.

Engagement is the active participation of the United States in relationships beyond our borders. It is, quite simply, the opposite of a self-imposed isolation that denies us the ability to shape outcomes. Indeed, America has never succeeded through isolationism. As the nation that helped to build our international system after World War II and to bring about the globalization that came with the end of the Cold War, we must reengage the world on a comprehensive and sustained basis.

Engagement begins with our closest friends and allies—from Europe to Asia; from North America to the Middle East. These nations share a common history of struggle on behalf of security, prosperity, and democracy. They share common values and a common commitment to international norms that recognize both the rights and responsibilities of all sovereign nations. America's national security depends on these vibrant alliances, and we must engage them as active partners in addressing global and regional security priorities and harnessing new opportunities to advance common interests. For instance, we pursue close and regular collaboration with our close allies the United Kingdom, France, and Germany on issues of mutual and global concern.

We will continue to deepen our cooperation with other 21st century centers of influence—including China, India, and Russia—on the basis of mutual interests and mutual respect. We will also pursue diplomacy and development that supports the emergence of new and successful partners, from the Americas to Africa; from the Middle East to Southeast Asia. Our ability to advance constructive cooperation is essential to the security and prosperity of specific regions, and to facilitating global cooperation on issues ranging from violent extremism and nuclear proliferation, to climate change, and global economic instability—issues that challenge all nations, but that no one nation alone can meet.

To adversarial governments, we offer a clear choice: abide by international norms, and achieve the political and economic benefits that come with greater integration with the international community; or refuse to accept this pathway, and bear the consequences of that decision, including greater isolation. Through engagement, we can create opportunities to resolve differences, strengthen the international community's support for our actions, learn about the intentions and nature of closed regimes, and plainly demonstrate to the publics within those nations that their governments are to blame for their isolation.

Successful engagement will depend upon the effective use and integration of different elements of American power. Our diplomacy and development capabilities must help prevent conflict, spur economic growth, strengthen weak and failing states, lift people out of poverty, combat climate change and epidemic disease, and strengthen institutions of democratic governance. Our military will continue strengthening its capacity to partner with foreign counterparts, train and assist security forces, and pursue military-to-military ties with a broad range of governments. We will continue to foster economic and financial transactions to advance our shared prosperity. And our intelligence and law enforcement agencies must cooperate effectively with foreign governments to anticipate events, respond to crises, and provide safety and security.

Finally, we will pursue engagement among peoples—not just governments—around the world. The United States Government will make a sustained effort to engage civil society and citizens and facilitate increased connections among the American people and peoples around the world—through efforts ranging from public service and educational exchanges, to increased commerce and private sector partnerships. In many instances, these modes of engagement have a powerful and enduring impact beyond our borders, and are a cost-effective way of projecting a positive vision of American leadership. Time and again, we have seen that the best ambassadors for American values and interests are the American people—our businesses, nongovernmental organizations, scientists, athletes, artists, military service members, and students.

Facilitating increased international engagement outside of government will help prepare our country to thrive in a global economy, while building the goodwill and relationships that are invaluable to sustaining American leadership. It also helps leverage strengths that are unique to America—our diversity and diaspora populations, our openness and creativity, and the values that our people embody in their own lives.

Promoting a Just and Sustainable International Order

Our engagement will underpin a just and sustainable international order—just, because it advances mutual interests, protects the rights of all, and holds accountable those who refuse to meet their responsibilities; sustainable because it is based on broadly shared norms and fosters collective action to address common challenges.

This engagement will pursue an international order that recognizes the rights and responsibilities of all nations. As we did after World War II, we must pursue a rules-based international system that can advance our own interests by serving mutual interests. International institutions must be more effective and representative of the diffusion of influence in the 21st century. Nations must have incentives to behave responsibly, or be isolated when they do not. The test of this international order must be the cooperation it facilitates and the results it generates—the ability of nations to come together to confront common challenges like violent extremism, nuclear proliferation, climate change, and a changing global economy.

That is precisely the reason we should strengthen enforcement of international law and our commitment to engage and modernize international institutions and frameworks. Those nations that refuse to meet their responsibilities will forsake the opportunities that come with international cooperation. Credible and effective alternatives to military action—from sanctions to isolation—must be strong enough to change behavior, just as we must reinforce our alliances and our military capabilities. And if nations challenge or undermine an international order that is based upon rights and responsibilities, they must find themselves isolated.

We succeeded in the post-World War II era by pursuing our interests within multilateral forums like the United Nations—not outside of them. We recognized that institutions that aggregated the national interests of many nations would never be perfect; but we also saw that they were an indispensable vehicle for pooling international resources and enforcing international norms. Indeed, the basis for international cooperation since World War II has been an architecture of international institutions, organizations, regimes, and standards that establishes certain rights and responsibilities for all sovereign nations.

In recent years America's frustration with international institutions has led us at times to engage the United Nations (U.N.) system on an ad hoc basis. But in a world of transnational challenges, the United States will need to invest in strengthening the international system, working from inside international institutions and frameworks to face their imperfections head on and to mobilize transnational cooperation.

We must be clear-eyed about the factors that have impeded effectiveness in the past. In order for collective action to be mobilized, the polarization that persists across region, race, and religion will need to be replaced by a galvanizing sense of shared interest. Swift and effective international action often turns on the political will of coalitions of countries that comprise regional or international institutions. New and emerging powers who seek greater voice and representation will need to accept greater responsibility for meeting global challenges. When nations breach agreed international norms, the countries who espouse those norms must be convinced to band together to enforce them.

We will expand our support to modernizing institutions and arrangements such as the evolution of the G-8 to the G-20 to reflect the realities of today's international environment. Working with the institutions and the countries that comprise them, we will enhance international capacity to prevent conflict, spur economic growth, improve security, combat climate change, and address the challenges posed by weak and failing states. And we will challenge and assist international institutions and frameworks to reform when they fail to live up to their promise. Strengthening the legitimacy and authority of international law and institutions, especially the U.N., will require a constant struggle to improve performance.

Furthermore, our international order must recognize the increasing influence of individuals in today's world. There must be opportunities for civil society to thrive within nations and to forge connections among them. And there must be opportunities for individuals and the private sector to play a major role in addressing common challenges—whether supporting a nuclear fuel bank, promoting global health, fostering entrepreneurship, or exposing violations of universal rights. In the 21st century, the ability of individuals and nongovernment actors to play a positive role in shaping the international environment represents a distinct opportunity for the United States.

Within this context, we know that an international order where every nation upholds its rights and responsibilities will remain elusive. Force will sometimes be necessary to confront threats. Technology will continue to bring with it new dangers. Poverty and disease will not be completely abolished. Oppression will always be with us. But if we recognize these challenges, embrace America's responsibility to confront them with its partners, and forge new cooperative approaches to get others to join us in overcoming them, then the international order of a globalized age can better advance our interests and the common interests of nations and peoples everywhere.

STRENGTHENING NATIONAL CAPACITY—A WHOLE OF GOVERNMENT APPROACH

To succeed, we must update, balance, and integrate all of the tools of American power and work with our allies and partners to do the same. Our military must maintain its conventional superiority and, as long as nuclear weapons exist, our nuclear deterrent capability, while continuing to enhance its capacity to defeat asymmetric threats, preserve access to the global commons, and strengthen partners. We must invest in diplomacy and development capabilities and institutions in a way that complements and reinforces our global partners. Our intelligence capabilities must continuously evolve to identify and characterize conventional and asymmetric threats and provide timely insight. And we must integrate our approach to homeland security with our broader national security approach.

We are improving the integration of skills and capabilities within our military and civilian institutions, so they complement each other and operate seamlessly. We are also improving coordinated planning and policymaking and must build our capacity in key areas where we fall short. This requires close cooperation with Congress and a deliberate and inclusive interagency process, so that we achieve integration of our efforts to implement and monitor operations, policies, and strategies. To initiate this effort, the White House merged the staffs of the National Security Council and Homeland Security Council.

However, work remains to foster coordination across departments and agencies. Key steps include more effectively ensuring alignment of resources with our national security strategy, adapting the education and training of national security professionals to equip them to meet modern challenges, reviewing authorities and mechanisms to implement and coordinate assistance programs, and other policies and programs that strengthen coordination.

- **Defense:** We are strengthening our military to ensure that it can prevail in today's wars; to prevent and deter threats against the United States, its interests, and our allies and partners; and prepare to defend the United States in a wide range of contingencies against state and nonstate actors. We will continue to rebalance our military capabilities to excel at counterterrorism, counterinsurgency, stability operations, and meeting increasingly sophisticated security threats, while ensuring our force is ready to address the full range of military operations. This includes preparing for increasingly sophisticated adversaries, deterring and defeating aggression in anti-access environments, and defending the United States and supporting civil authorities at home. The most valuable component of our national defense is the men and women who make up America's all-volunteer force. They have shown tremendous resilience, adaptability, and capacity for innovation, and we will provide our service members with the resources that they need to succeed and rededicate ourselves to providing support and care for wounded warriors, veterans, and military families. We must set the force on a path to sustainable deployment cycles and preserve and enhance the long-term viability of our force through successful recruitment, retention, and recognition of those who serve.
- **Diplomacy:** Diplomacy is as fundamental to our national security as our defense capability. Our diplomats are the first line of engagement, listening to our partners, learning from them, building respect for one another, and seeking common ground. Diplomats, development experts, and others in the United States Government must be able to work side by side to support a common agenda. New skills are needed to foster effective interaction to convene, connect, and mobilize not only other governments and international organizations, but also nonstate actors such as corporations, foundations, nongovernmental organizations, universities, think tanks, and faith-based organizations, all of whom increasingly have a distinct role to play on both diplomatic and development issues. To accomplish these goals our diplomatic personnel and missions must be expanded at home and abroad to support the increasingly transnational nature of 21st century security challenges. And we must provide the appropriate authorities and mechanisms to implement and coordinate assistance programs and

grow the civilian expeditionary capacity required to assist governments on a diverse array of issues.

- **Economic:** Our economic institutions are crucial components of our national capacity and our economic instruments are the bedrock of sustainable national growth, prosperity and influence. The Office of Management and Budget, Departments of the Treasury, State, Commerce, Energy, and Agriculture, United States Trade Representative, Federal Reserve Board, and other institutions help manage our currency, trade, foreign investment, deficit, inflation, productivity, and national competitiveness. Remaining a vibrant 21st century economic power also requires close cooperation between and among developed nations and emerging markets because of the interdependent nature of the global economy. America—like other nations—is dependent upon overseas markets to sell its exports and maintain access to scarce commodities and resources. Thus, finding overlapping mutual economic interests with other nations and maintaining those economic relationships are key elements of our national security strategy.
- **Development:** Development is a strategic, economic, and moral imperative. We are focusing on assisting developing countries and their people to manage security threats, reap the benefits of global economic expansion, and set in place accountable and democratic institutions that serve basic human needs. Through an aggressive and affirmative development agenda and commensurate resources, we can strengthen the regional partners we need to help us stop conflicts and counter global criminal networks; build a stable, inclusive global economy with new sources of prosperity; advance democracy and human rights; and ultimately position ourselves to better address key global challenges by growing the ranks of prosperous, capable, and democratic states that can be our partners in the decades ahead. To do this, we are expanding our civilian development capability; engaging with international financial institutions that leverage our resources and advance our objectives; pursuing a development budget that more deliberately reflects our policies and our strategy, not sector earmarks; and ensuring that our policy instruments are aligned in support of development objectives.
- **Homeland Security:** Homeland security traces its roots to traditional and historic functions of government and society, such as civil defense, emergency response, law enforcement, customs, border patrol, and

immigration. In the aftermath of 9/11 and the foundation of the Department of Homeland Security, these functions have taken on new organization and urgency. Homeland security, therefore, strives to adapt these traditional functions to confront new threats and evolving hazards. It is not simply about government action alone, but rather about the collective strength of the entire country. Our approach relies on our shared efforts to identify and interdict threats; deny hostile actors the ability to operate within our borders; maintain effective control of our physical borders; safeguard lawful trade and travel into and out of the United States; disrupt and dismantle transnational terrorist, and criminal organizations; and ensure our national resilience in the face of the threat and hazards. Taken together, these efforts must support a homeland that is safe and secure from terrorism and other hazards and in which American interests, aspirations, and way of life can thrive.

- **Intelligence:** Our country's safety and prosperity depend on the quality of the intelligence we collect and the analysis we produce, our ability to evaluate and share this information in a timely manner, and our ability to counter intelligence threats. This is as true for the strategic intelligence that informs executive decisions as it is for intelligence support to homeland security, state, local, and tribal governments, our troops, and critical national missions. We are working to better integrate the Intelligence Community, while also enhancing the capabilities of our Intelligence Community members. We are strengthening our partnerships with foreign intelligence services and sustaining strong ties with our close allies. And we continue to invest in the men and women of the Intelligence Community.
- **Strategic Communications:** Across all of our efforts, effective strategic communications are essential to sustaining global legitimacy and supporting our policy aims. Aligning our actions with our words is a shared responsibility that must be fostered by a culture of communication throughout government. We must also be more effective in our deliberate communication and engagement and do a better job understanding the attitudes, opinions, grievances, and concerns of peoples—not just elites—around the world. Doing so allows us to convey credible, consistent messages and to develop effective plans, while better understanding how our actions will be

perceived. We must also use a broad range of methods for communicating with foreign publics, including new media.

- **The American People and the Private Sector:** The ideas, values, energy, creativity, and resilience of our citizens are America's greatest resource. We will support the development of prepared, vigilant, and engaged communities and underscore that our citizens are the heart of a resilient country. And we must tap the ingenuity outside government through strategic partnerships with the private sector, nongovernmental organizations, foundations, and community-based organizations. Such partnerships are critical to U.S. success at home and abroad, and we will support them through enhanced opportunities for engagement, coordination, transparency, and information sharing.

III. Advancing Our Interests

To achieve the world we seek, the United States must apply our strategic approach in pursuit of four enduring national interests:

- **Security:** The security of the United States, its citizens, and U.S. allies and partners.
- **Prosperity:** A strong, innovative, and growing U.S. economy in an open international economic system that promotes opportunity and prosperity.
- **Values:** Respect for universal values at home and around the world.
- **International Order:** An international order advanced by U.S. leadership that promotes peace, security, and opportunity through stronger cooperation to meet global challenges.

Each of these interests is inextricably linked to the others: no single interest can be pursued in isolation, but at the same time, positive action in one area will help advance all four. The initiatives described below do not encompass all of America's national security concerns. However, they represent areas of particular priority and areas where progress is critical to securing our country and renewing American leadership in the years to come.

Security

> "We will not apologize for our way of life, nor will we waver in its defense. And for those who seek to advance their aims by inducing terror and slaughtering innocents, we say to you now that our spirit is stronger and cannot be broken—you cannot outlast us, and we will defeat you."
>
> —President Barack Obama, Inaugural Address, January 20, 2009

The threats to our people, our homeland, and our interests have shifted dramatically in the last 20 years. Competition among states endures, but instead of a single nuclear adversary, the United States is now threatened by the potential spread of nuclear weapons to extremists who may not be deterred from using them. Instead of a hostile expansionist empire, we now face a diverse array of challenges, from a loose network of violent extremists to states that flout international norms or face internal collapse. In addition to facing enemies on traditional battlefields, the United States must now be prepared for asymmetric threats, such as those that target our reliance on space and cyberspace.

This Administration has no greater responsibility than protecting the American people. Furthermore, we embrace America's unique responsibility to promote international security—a responsibility that flows from our commitments to allies, our leading role in supporting a just and sustainable international order, and our unmatched military capabilities.

The United States remains the only nation able to project and sustain large-scale military operations over extended distances. We maintain superior capabilities to deter and defeat adaptive enemies and to ensure the credibility of security partnerships that are fundamental to regional and global security. In this way, our military continues to underpin our national security and global leadership, and when we use it appropriately, our security and leadership is reinforced. But when we overuse our military might, or fail to invest in or deploy complementary tools, or act without partners, then our military is overstretched, Americans bear a greater burden, and our leadership around the world is too narrowly identified with military force. And we know that our enemies aim to overextend our Armed Forces and to drive wedges between us and those who share our interests.

Therefore, we must continue to adapt and rebalance our instruments of statecraft. At home, we are integrating our homeland security efforts seamlessly with other aspects of our national security approach, and strengthening our preparedness and resilience. Abroad, we are strengthening alliances, forging new partnerships, and using every tool of American power to advance our objectives—including enhanced diplomatic and development capabilities with the ability both to prevent conflict and to work alongside our military. We are strengthening international norms to isolate governments that flout them and to marshal cooperation against nongovernmental actors who endanger our common security.

Strengthen Security and Resilience at Home

At home, the United States is pursuing a strategy capable of meeting the full range of threats and hazards to our communities. These threats and hazards include terrorism, natural disasters, large-scale cyber attacks, and pandemics. As we do everything within our power to prevent these dangers, we also recognize that we will not be able to deter or prevent every single threat. That is why we must also enhance our resilience—the ability to adapt to changing conditions and prepare for, withstand, and rapidly recover from disruption. To keep Americans safe and secure at home, we are working to:

Enhance Security at Home: Security at home relies on our shared efforts to prevent and deter attacks by identifying and interdicting threats, denying hostile actors the ability to operate within our borders, protecting the nation's critical infrastructure and key resources, and securing cyberspace. That is why we are pursuing initiatives to protect and reduce vulnerabilities in critical infrastructure, at our borders, ports, and airports, and to enhance overall air, maritime, transportation, and space and cyber security. Building on this foundation, we recognize that the global systems that carry people, goods, and data around the globe also facilitate the movement of dangerous people, goods, and data. Within these systems of transportation and transaction, there are key nodes—for example, points of origin and transfer, or border crossings—that represent opportunities for exploitation and interdiction. Thus, we are working with partners abroad to confront threats that often begin beyond our borders. And we are developing lines of coordination at home across Federal, state, local, tribal, territorial, nongovernmental, and private-sector partners, as well as individuals and communities.

Effectively Manage Emergencies: We are building our capability to prepare for disasters to reduce or eliminate long-term effects to people and their property from hazards and to respond to and recover from major incidents. To improve our preparedness, we are integrating domestic all hazards planning at all levels of government and building key capabilities to respond to emergencies. We continue to collaborate with communities to ensure preparedness efforts are integrated at all levels of government with the private and nonprofit sectors. We are investing in operational capabilities and equipment, and improving the reliability and interoperability of communications systems for first responders. We are encouraging domestic regional planning and integrated preparedness programs and will encourage government at all levels to engage in long-term recovery planning. It is critical that we continually test and improve plans using exercises that are realistic in scenario and consequences.

Empowering Communities to Counter Radicalization: Several recent incidences of violent extremists in the United States who are committed to fighting here and abroad have underscored the threat to the United States and our interests posed by individuals radicalized at home. Our best defenses against this threat are well informed and equipped families, local communities, and institutions. The Federal Government will invest in intelligence to understand this threat and expand community engagement and development programs to empower local communities. And the Federal Government, drawing on the expertise and resources from all relevant agencies, will clearly communicate our policies and intentions, listening to local concerns, tailoring policies to address regional concerns, and making clear that our diversity is part of our strength—not a source of division or insecurity.

Improve Resilience Through Increased Public-Private Partnerships: When incidents occur, we must show resilience by maintaining critical operations and functions, returning to our normal life, and learning from disasters so that their lessons can be translated into pragmatic changes when necessary. The private sector, which owns and operates most of the nation's critical infrastructure, plays a vital role in preparing for and recovering from disasters. We must, therefore, strengthen public-private partnerships by developing incentives for government and the private sector to design structures and systems that can withstand disruptions and mitigate associated consequences, ensure redundant systems where necessary to maintain the ability to operate, decentralize critical operations to reduce our vulnerability to single points of disruption, develop and test continuity plans to ensure the ability to restore critical capabilities, and invest in improvements and maintenance of existing infrastructure.

Engage with Communities and Citizens: We will emphasize individual and community preparedness and resilience through frequent engagement that provides clear and reliable risk and emergency information to the public. A key part of this effort is providing practical steps that all Americans can take to protect themselves, their families, and their neighbors. This includes transmitting information through multiple pathways and to those with special needs. In addition, we support efforts to develop a nationwide public safety broadband network. Our efforts to inform and empower Americans and their communities recognize that resilience has always been at the heart of the American spirit.

Disrupt, Dismantle, and Defeat Al-Qa'ida and its Violent Extremist Affiliates in Afghanistan, Pakistan, and around the World

The United States is waging a global campaign against al-Qa'ida and its terrorist affiliates. To disrupt, dismantle and defeat al-Qa'ida and its affiliates, we are pursuing a strategy that protects our homeland, secures the world's most dangerous weapons and material, denies al-Qa'ida safe haven, and builds positive partnerships with Muslim communities around the world. Success requires a broad, sustained, and integrated campaign that judiciously applies every tool of American power—both military and civilian—as well as the concerted efforts of like-minded states and multilateral institutions.

We will always seek to delegitimize the use of terrorism and to isolate those who carry it out. Yet this is not a global war against a tactic—terrorism or a religion—Islam. We are at war with a specific network, al-Qa'ida, and its terrorist affiliates who support efforts to attack the United States, our allies, and partners.

Prevent Attacks on and in the Homeland: To prevent acts of terrorism on American soil, we must enlist all of our intelligence, law enforcement, and homeland security capabilities. We will continue to integrate and leverage state and major urban area fusion centers that have the capability to share classified information; establish a nationwide framework for reporting suspicious activity; and implement an integrated approach to our counterterrorism information systems to ensure that the analysts, agents, and officers who protect us have access to all relevant intelligence throughout the government. We are improving information sharing and cooperation by linking networks to facilitate Federal, state, and local capabilities to seamlessly exchange messages and information, conduct searches, and collaborate. We are coordinating better with foreign partners to identify, track, limit access to funding, and prevent terrorist travel. Recognizing the inextricable link between domestic and transnational security, we will collaborate bilaterally, regionally, and through international institutions to promote global efforts to prevent terrorist attacks.

Strengthen Aviation Security: We know that the aviation system has been a particular target of al-Qa'ida and its affiliates. We must continue to bolster aviation security worldwide through a focus on increased information collection and sharing, stronger passenger vetting and screening measures, the development and development of advanced screening technologies, and cooperation with the international community to strengthen aviation security standards and efforts around the world.

Deny Terrorists Weapons of Mass Destruction: To prevent acts of terrorism with the world's most dangerous weapons, we are dramatically accelerating and intensifying efforts to secure all vulnerable nuclear materials by the end of 2013, and to prevent the spread of nuclear weapons. We will also take actions to safeguard knowledge and capabilities in the life and chemical sciences that could be vulnerable to misuse.

Deny Al-Qa'ida the Ability to Threaten the American People, Our Allies, Our Partners and Our Interests Overseas: Al-Qa'ida and its allies must not be permitted to gain or retain any capacity to plan and launch international terrorist attacks, especially against the U.S. homeland. Al Qa'ida's core in Pakistan remains the most dangerous component of the larger network, but we also face a growing threat from the group's allies worldwide. We must deny these groups the ability to conduct operational plotting from any locale, or to recruit, train, and position operatives, including those from Europe and North America.

Afghanistan and Pakistan: This is the epicenter of the violent extremism practiced by al Qa'ida. The danger from this region will only grow if its security slides backward, the Taliban controls large swaths of Afghanistan, and al-Qa'ida is allowed to operate with impunity. To prevent future attacks on the United States, our allies, and partners, we must work with others to keep the pressure on al-Qa'ida and increase the security and capacity of our partners in this region.

In Afghanistan, we must deny al-Qa'ida a safe haven, deny the Taliban the ability to overthrow the government, and strengthen the capacity of

Afghanistan's security forces and government so that they can take lead responsibility for Afghanistan's future. Within Pakistan, we are working with the government to address the local, regional, and global threat from violent extremists.

We will achieve these objectives with a strategy comprised of three components.

- First, our military and International Security Assistance Force (ISAF) partners within Afghanistan are targeting the insurgency, working to secure key population centers, and increasing efforts to train Afghan security forces. These military resources will allow us to create the conditions to transition to Afghan responsibility. In July 2011, we will begin reducing our troops responsibly, taking into account conditions on the ground. We will continue to advise and assist Afghanistan's Security Forces so that they can succeed over the long term.
- Second, we will continue to work with our partners, the United Nations, and the Afghan Government to improve accountable and effective governance. As we work to advance our strategic partnership with the Afghan Government, we are focusing assistance on supporting the President of Afghanistan and those ministries, governors, and local leaders who combat corruption and deliver for the people. Our efforts will be based upon performance, and we will measure progress. We will also target our assistance to areas that can make an immediate and enduring impact in the lives of the Afghan people, such as agriculture, while supporting the human rights of all of Afghanistan's people—women and men. This will support our long-term commitment to a relationship between our two countries that supports a strong, stable, and prosperous Afghanistan.
- Third, we will foster a relationship with Pakistan founded upon mutual interests and mutual respect. To defeat violent extremists who threaten both of our countries, we will strengthen Pakistan's capacity to target violent extremists within its borders, and continue to provide security assistance to support those efforts. To strengthen Pakistan's democracy and development, we will provide substantial assistance responsive to the needs of the Pakistani people, and sustain a long-term partnership committed to Pakistan's future. The strategic partnership that we are developing with Pakistan includes deepening cooperation in a broad range of areas, addressing both security and

civilian challenges, and we will continue to expand those ties through our engagement with Pakistan in the years to come.

Deny Safe Havens and Strengthen At-Risk States: Wherever al-Qa'ida or its terrorist affiliates attempt to establish a safe haven—as they have in Yemen, Somalia, the Maghreb, and the Sahel—we will meet them with growing pressure. We also will strengthen our own network of partners to disable al-Qa'ida's financial, human, and planning networks; disrupt terrorist operations before they mature; and address potential safe-havens before al-Qa'ida and its terrorist affiliates can take root. These efforts will focus on information-sharing, law enforcement cooperation, and establishing new practices to counter evolving adversaries. We will also help states avoid becoming terrorist safe havens by helping them build their capacity for responsible governance and security through development and security sector assistance.

Deliver Swift and Sure Justice: To effectively detain, interrogate, and prosecute terrorists, we need durable legal approaches consistent with our security and our values. We adhere to several principles: we will leverage all available information and intelligence to disrupt attacks and dismantle al-Qa'ida and affiliated terrorist organizations; we will bring terrorists to justice; we will act in line with the rule of law and due process; we will submit decisions to checks and balances and accountability; and we will insist that matters of detention and secrecy are addressed in a manner consistent with our Constitution and laws. To deny violent extremists one of their most potent recruitment tools, we will close the prison at Guantanamo Bay.

Resist Fear and Overreaction: The goal of those who perpetrate terrorist attacks is in part to sow fear. If we respond with fear, we allow violent extremists to succeed far beyond the initial impact of their attacks, or attempted attacks—altering our society and enlarging the standing of al-Qa'ida and its terrorist affiliates far beyond its actual reach. Similarly, overreacting in a way that creates fissures between America and certain regions or religions will undercut our leadership and make us less safe.

Contrast Al-Qa'ida's Intent to Destroy with Our Constructive Vision: While violent extremists seek to destroy, we will make clear our intent to build. We are striving to build bridges among people of different faiths and regions. We will continue to work to resolve the Arab-Israeli conflict, which has long been a source of tension. We will continue to stand up for the universal rights of all people, even for those with whom we disagree. We are developing new partnerships in Muslim communities around the world on behalf of health, education, science, employment, and innovation. And through our broader emphasis on Muslim engagement, we will communicate our commitment to support the aspirations of all people for security and opportunity.

USE OF FORCE

Military force, at times, may be necessary to defend our country and allies or to preserve broader peace and security, including by protecting civilians facing a grave humanitarian crisis. We will draw on diplomacy, development, and international norms and institutions to help resolve disagreements, prevent conflict, and maintain peace, mitigating where possible the need for the use of force. This means credibly underwriting U.S. defense commitments with tailored approaches to deterrence and ensuring the U.S. military continues to have the necessary capabilities across all domains—land, air, sea, space, and cyber. It also includes helping our allies and partners build capacity to fulfill their responsibilities to contribute to regional and global security.

While the use of force is sometimes necessary, we will exhaust other options before war whenever we can, and carefully weigh the costs and risks of action against the costs and risks of inaction. When force is necessary, we will continue to do so in a way that reflects our values and strengthens our legitimacy, and we will seek broad international support, working with such institutions as NATO and the U.N. Security Council.

The United States must reserve the right to act unilaterally if necessary to defend our nation and our interests, yet we will also seek to adhere to standards that govern the use of force. Doing so strengthens those who act in line with international standards, while isolating and weakening those who do not. We will also outline a clear mandate and specific objectives and thoroughly consider the consequences —intended and unintended—of our actions. And the United States will take care when sending the men and women of our Armed Forces into harm's way to ensure they have the leadership, training, and equipment they require to accomplish their mission.

Finally, we reject the notion that al-Qa'ida represents any religious authority. They are not religious leaders, they are killers; and neither Islam nor any other religion condones the slaughter of innocents.

Reverse the Spread of Nuclear and Biological Weapons and Secure Nuclear Materials

The American people face no greater or more urgent danger than a terrorist attack with a nuclear weapon. And international peace and security is threatened by proliferation that could lead to a nuclear exchange. Indeed, since the end of the Cold War, the risk of a nuclear attack has increased. Excessive Cold War stockpiles remain. More nations have acquired nuclear weapons. Testing has continued. Black markets trade in nuclear secrets and materials. Terrorists are determined to buy, build, or steal a nuclear weapon. Our efforts to contain these dangers are centered in a global nonproliferation regime that has frayed as more people and nations break the rules.

That is why reversing the spread of nuclear weapons is a top priority. Success depends upon broad consensus and concerted action, we will move forward strategically on a number of fronts through our example, our partnerships, and a reinvigorated international regime. The United States will:

Pursue the Goal of a World Without Nuclear Weapons: While this goal will not be reached during this Administration, its active pursuit and eventual achievement will increase global security, keep our commitment under the NPT, build our cooperation with Russia and other states, and increase our credi bility to hold others accountable for their obligations. As long as any nuclear weapons exist, the United States will sustain a safe, secure, and effective nuclear arsenal, both to deter potential adversaries and to assure U.S. allies and other security partners that they can count on America's security commitments. But we have signed and seek to ratify a landmark New START Treaty with Russia to substantially limit our deployed nuclear warheads and strategic delivery vehicles, while assuring a comprehensive monitoring regime. We are reducing the role of nuclear weapons in our national security approach, extending a negative security assurance not to use or threaten to use nuclear weapons against those nonnuclear nations that are in compliance with the NPT and their nuclear nonproliferation obligations, and investing in the modernization of a safe, secure, and effective stockpile without the production of new nuclear weapons. We will pursue ratification of the Comprehensive Test Ban Treaty. And we will seek a new treaty that verifiably ends the production of fissile materials intended for use in nuclear weapons.

Strengthen the Nuclear Non-Proliferation Treaty: The basic bargain of the NPT is sound: countries with nuclear weapons will move toward disarmament; countries without nuclear weapons will forsake them; and all countries can access peaceful nuclear energy. To strengthen the NPT, we will seek more resources and authority for international inspections. We will develop a new framework for civil nuclear cooperation. As members of the Global Nuclear Energy Partnership have agreed, one important element of an enhanced framework could be cradle-to-grave nuclear fuel management. We will pursue a broad, international consensus to insist that all nations meet their obligations. And we will also pursue meaningful consequences for countries that fail to meet their obligations under the N PT or to meet the requirements for withdrawing from it.

Present a Clear Choice to Iran and North Korea: The United States will pursue the denuclearization of the Korean peninsula and work to prevent Iran from developing a nuclear weapon. This is not about singling out nations—it is about the responsibilities of all nations and the success of the nonproliferation regime. Both nations face a clear choice. If North Korea eliminates its nuclear weapons program, and Iran meets its international obligations on its nuclear program, they will be able to proceed on a path to greater political and economic integration with the international community. If they ignore their international obligations, we will pursue multiple means to increase their isolation and bring them into compliance with international nonproliferation norms.

Secure Vulnerable Nuclear Weapons and Material: The Global Nuclear Security Summit of 2010 rallied 47 nations behind the goal of securing all nuclear materials from terrorist groups. By the end of 2013, we will seek to complete a focused international effort to secure all vulnerable nuclear material around the world through enhanced protection and accounting practices, expanded cooperation with and through international institutions, and new partnerships to lock down these sensitive materials. To detect and intercept nuclear materials in transit, and to stop the illicit trade in these technologies, we will work to turn programs such as the Proliferation Security Initiative and the Global Initiative to Combat Nuclear Terrorism into durable international efforts. And we will sustain broad-based cooperation with other nations and international institutions to ensure the continued improvements necessary to protect nuclear materials from evolving threats.

Support Peaceful Nuclear Energy: As countries move increasingly to tap peaceful nuclear energy to provide power generation while advancing climate goals, the world must develop an infrastructure in the countries that seek to use nuclear energy for their energy security needs and climate goals to ensure that nuclear energy is developed in a safer manner. We will do so by promoting safety through regulatory bodies and training of operators, promoting physical security to prevent terrorist acts, and assuring safe and secure handling of fuel at the front and back ends of the nuclear fuel cycle.

Counter Biological Threats: The effective dissemination of a lethal biological agent within a population center would endanger the lives of hundreds of thousands of people and have unprecedented economic, societal, and political consequences. We must continue to work at home with first responders and health officials to reduce the risk associated with unintentional or deliberate outbreaks of infectious disease and to strengthen our resilience across the spectrum of high-consequence biological threats. We will work with domestic and international partners to protect against biological threats by promoting global health security and reinforcing norms of safe and responsible conduct; obtaining timely and accurate insight on current and emerging risks; taking reasonable steps to reduce the potential for exploitation; expanding our capability to prevent, attribute, and apprehend those who carry out attacks; communicating effectively with all stakeholders; and helping to transform the international dialogue on biological threats.

Advance Peace, Security, and Opportunity in the Greater Middle East

The United States has important interests in the greater Middle East. They include broad cooperation on a wide range of issues with our close friend, Israel, and an unshakable commitment to its security; the achievement of the Palestinian people's legitimate aspirations for statehood, opportunity, and the realization of their extraordinary potential; the unity and security of Iraq and the fostering of its democracy and reintegration into the region; the transformation of Iranian policy away from its pursuit of nuclear weapons, support for terrorism, and threats against its neighbors; nonproliferation; and counterterrorism cooperation, access to energy, and integration of the region into global markets.

At the same time, our engagement must be both comprehensive and strategic. It should extend beyond near-term threats by appealing to peoples' aspirations for justice, education, and opportunity and by pursuing a positive and sustainable vision of U.S. partnership with the region. Furthermore, our relationship with our Israeli and Arab friends and partners in the region extends beyond our commitment to its security and includes the continued ties we share in areas such as trade, exchanges, and cooperation on a broad range of issues.

Complete a Responsible Transition as We End the War in Iraq: The war in Iraq presents a distinct and important challenge to the United States, the international community, the Iraqi people, and the region. America's servicemen and women, along with our coalition partners, have performed remarkably in fighting determined enemies and have worked with our civilians to help the Iraqi people regain control of their own destiny. Going forward, we have a responsibility, for our own security and the security of the region, to successfully end the war through a full transition to Iraqi responsibility. We will cultivate an enduring relationship with Iraq based on mutual interests and mutual respect.

Our goal is an Iraq that is sovereign, stable, and self-reliant. To achieve that goal, we are continuing to promote an Iraqi Government that is just, representative, and accountable and that denies support and safe haven to terrorists. The United States will pursue no claim on Iraqi territory or resources, and we will keep our commitments to Iraq's democratically elected government. These efforts will build new ties of trade and commerce between Iraq and the world, enable Iraq to assume its rightful place in the community of nations, and contribute to the peace and security of the region.

We are pursuing these objectives with a strategy that has three core components.

- **Transition Security:** First, we are transitioning security to full Iraqi responsibility. We will end the combat mission in Iraq by the end of August 2010. We will continue to train, equip, and advise Iraqi Security Forces; conduct targeted counterterrorism missions; and protect ongoing civilian and military efforts in Iraq. And, consistent with our commitments to the Iraqi Government, including the U.S.-Iraq Security Agreement, we will remove all of our troops from Iraq by the end of 2011.

- **Civilian Support:** Second, as the security situation continues to improve, U.S. civilian engagement will deepen and broaden. We will sustain a capable political, diplomatic, and civilian effort to help the Iraqi people as they resolve outstanding differences, integrate those refugees and displaced persons who can return, and continue to develop accountable democratic institutions that can better serve their basic needs. We will work with our Iraqi partners to implement the Strategic Framework Agreement, with the Department of State taking the lead. This will include cooperation on a range of issues including defense and security cooperation, political and diplomatic cooperation, rule of law, science, health, education, and economics.
- **Regional Diplomacy and Development:** Third, we will continue to pursue comprehensive engagement across the region to ensure that our drawdown in Iraq provides an opportunity to advance lasting security and sustainable development for both Iraq and the broader Middle East. The United States will continue to retain a robust civilian presence commensurate with our strategic interests in the country and the region. We are transforming our relationship to one consistent with other strategic partners in the region.

Pursue Arab-Israeli Peace: The United States, Israel, the Palestinians, and the Arab States have an interest in a peaceful resolution of the Arab-Israeli conflict—one in which the legitimate aspirations of Israelis and Palestinians for security and dignity are realized, and Israel achieves a secure and lasting peace with all of its neighbors.

The United States seeks two states living side by side in peace and security—a Jewish state of Israel, with true security, acceptance, and rights for all Israelis; and a viable, independent Palestine with contiguous territory that ends the occupation that began in 1967 and realizes the potential of the Palestinian people. We will continue to work regionally and with like-minded partners in order to advance negotiations that address the permanent-status issues: security for Israelis and Palestinians; borders, refugees, and Jerusalem. We also seek international support to build the institutions upon which a Palestinian state will depend, while supporting economic development that can bring opportunity to its people.

Any Arab-Israeli peace will only be lasting if harmful regional interference ends and constructive regional support deepens. As we pursue peace between Israelis and Palestinians, we will also pursue peace between Israel and Lebanon, Israel and Syria, and a broader peace between Israel and its neighbors. We will pursue regional initiatives with multilateral participation, alongside bilateral negotiations.

Promote a Responsible Iran: For decades, the Islamic Republic of Iran has endangered the security of the region and the United States and failed to live up to its international responsibilities. In addition to its illicit nuclear program, it continues to support terrorism, undermine peace between Israelis and Palestinians, and deny its people their universal rights. Many years of refusing to engage Iran failed to reverse these trends; on the contrary, Iran's behavior became more threatening. Engagement is something we pursue without illusion. It can offer Iran a pathway to a better future, provided Iran's leaders are prepared to take it. But that better pathway can only be achieved if Iran's leaders change course, act to restore the confidence of the international community, and fulfill their obligations. The United States seeks a future in which Iran meets its international responsibilities, takes its rightful place in the community of nations, and enjoys the political and economic opportunities that its people deserve. Yet if the Iranian Government continues to refuse to live up to its international obligations, it will face greater isolation.

Invest in the Capacity of Strong and Capable Partners

Where governments are incapable of meeting their citizens' basic needs and fulfilling their responsibilities to provide security within their borders, the consequences are often global and may directly threaten the American people. To advance our common security, we must address the underlying political and economic deficits that foster instability, enable radicalization and extremism, and ultimately undermine the ability of governments to manage threats within their borders and to be our partners in addressing common challenges. To invest in the capacity of strong and capable partners, we will work to:

Foster Security and Reconstruction in the Aftermath of Conflict: The United States and the international community cannot shy away from the difficult task of pursuing stabilization in conflict and post-conflict environments. In countries like Iraq and Afghanistan, building the capacity necessary for security, economic growth, and good governance is the only path to long term peace and security. But we have also learned that the effectiveness of these efforts is profoundly affected by the capacity of governments and the political will of their leaders. We will take these constraints into account in designing appropriate assistance strategies and will facilitate the kind of collaboration that is essential—within our government and with international organizations—in those instances when we engage in the difficult work of helping to bring conflicts to an end.

Pursue Sustainable and Responsible Security Systems in At-Risk States: Proactively investing in stronger societies and human welfare is far more effective and efficient than responding after state collapse. The United States must improve its capability to strengthen the security of states at risk of conflict and violence. We will undertake long-term, sustained efforts to strengthen the capacity of security forces to guarantee internal security, defend against external threats, and promote regional security and respect for human rights and the rule of law. We will also continue to strengthen the administrative and oversight capability of civilian security sector institutions, and the effectiveness of criminal justice.

Prevent the Emergence of Conflict: Our strategy goes beyond meeting the challenges of today, and includes preventing the challenges and seizing the opportunities of tomorrow. This requires investing now in the capable partners of the future; building today the capacity to strengthen the foundations of our common security, and modernizing our capabilities in order to ensure that we are agile in the face of change. We have already begun to reorient and strengthen our development agenda; to take stock of and enhance our capabilities; and to forge new and more effective means of applying the skills of our military, diplomats, and development experts. These kinds of measures will help us diminish military risk, act before crises and conflicts erupt, and ensure that governments are better able to serve their people.

Secure Cyberspace

Cybersecurity threats represent one of the most serious national security, public safety, and economic challenges we face as a nation. The very technologies that empower us to lead and create also empower those who would disrupt and destroy. They enable our military superiority, but our unclassified government networks are constantly probed by intruders. Our daily lives and public safety depend on power and electric grids, but potential adversaries could use cyber vulnerabilities to disrupt them on a massive scale. The Internet and e-commerce are keys to our economic competitiveness, but cyber criminals have cost companies and consumers hundreds of millions of dollars and valuable intellectual property.

The threats we face range from individual criminal hackers to organized criminal groups, from terrorist networks to advanced nation states. Defending against these threats to our security, prosperity, and personal privacy requires networks that are secure, trustworthy, and resilient. Our digital infrastructure, therefore, is a strategic national asset, and protecting it—while safeguarding privacy and civil liberties—is a national security priority. We will deter, prevent, detect, defend against, and quickly recover from cyber intrusions and attacks by:

Investing in People and Technology: To advance that goal, we are working across the government and with the private sector to design more secure technology that gives us the ability to better protect and to improve the resilience of critical government and industry systems and networks. We will continue to invest in the cutting-edge research and development necessary for the innovation and discovery we need to meet these challenges. We have begun a comprehensive national campaign to promote cybersecurity awareness and digital literacy from our boardrooms to our classrooms and to build a digital workforce for the 21st century.

Strengthening Partnerships: Neither government nor the private sector nor individual citizens can meet this challenge alone—we will expand the ways we work together. We will also strengthen our international partnerships on a range of issues, including the development of norms for acceptable conduct in cyberspace; laws concerning cybercrime; data preservation, protection, and privacy; and approaches for network defense and response to cyber attacks. We will work with all the key players— including all levels of government and the private sector, nationally and internationally—to investigate cyber intrusion and to ensure an organized and unified response to future cyber incidents. Just as we do for natural disasters, we have to have plans and resources in place beforehand.

Prosperity

> "The answers to our problems don't lie beyond our reach. They exist in our laboratories and universities; in our fields and our factories; in the imaginations of our entrepreneurs and the pride of the hardest-working people on Earth. Those qualities that have made America the greatest force of progress and prosperity in human history we still possess in ample measure. What is required now is for this country to pull together, confront boldly the challenges we face, and take responsibility for our future once more."

—President Barack Obama, Address to Joint Session of Congress, February 24, 2009

The foundation of American leadership must be a prosperous American economy. And a growing and open global economy serves as a source of opportunity for the American people and a source of strength for the United States. The free flow of information, people, goods, and services has also advanced peace among nations, as those places that have emerged more prosperous are often more stable. Yet we have also seen how shocks to the global economy can precipitate disaster—including the loss of jobs, a decline in standards of living in parts of our country, and instability and a loss of U.S. influence abroad. Meanwhile, growing prosperity around the world has made economic power more diffuse, creating a more competitive environment for America's people and businesses.

To allow each American to pursue the opportunity upon which our prosperity depends, we must build a stronger foundation for economic growth. That foundation must include access to a complete and competitive education for every American; a transformation of the way that we produce and use energy, so that we reduce our dependence on fossil fuels and lead the world in creating new jobs and industry; access to quality, affordable health care so our people, businesses, and government are not constrained by rising costs; and the responsible management of our Federal budget so that we balance our priorities and are not burdened by debt. To succeed, we must also ensure that America stays on the cutting edge of the science and innovation that supports our prosperity, defense, and international technological leadership.

This new foundation must underpin and sustain an international economic system that is critical to both our prosperity and to the peace and security of the world. We must reinvigorate and fortify it for the 21st century: by preventing cycles of boom and bust with new rules of the road at home and abroad; by saving more and spending less; by resisting protectionism and promoting trade that is free and fair; by coordinating our actions with other countries, and reforming international institutions to give emerging economies a greater voice and greater responsibility; and by supporting development that promotes good governance, unleashes the potential of different populations, and creates new markets overseas. Taken together, these actions can ensure inclusive growth that is balanced and sustained.

Strengthen Education and Human Capital

In a global economy of vastly increased mobility and interdependence, our own prosperity and leadership depends increasingly on our ability to provide our citizens with the education that they need to succeed, while attracting the premier human capital for our workforce. We must ensure that the most innovative ideas take root in America, while providing our people with the skills that they need to compete. That means we must:

Improve Education at All Levels: The United States has lost ground in education, even as our competitiveness depends on educating our children to succeed in a global economy based on knowledge and innovation. We are working to provide a complete and competitive education for all Americans, to include supporting high standards for early learning, reforming public schools, increasing access to higher education and job training, and promoting high-demand skills and education for emerging industries. We will also restore U.S. leadership in higher education by seeking the goal of leading the world in the proportion of college graduates by 2020.

Invest in Science, Technology, Engineering, and Math Education (STEM): America's long-term leadership depends on educating and producing future scientists and innovators. We will invest more in STEM education so students can learn to think critically in science, math, engineering, and technology; improve the quality of math and science teaching so American students are no longer outperformed by those in other nations; and expand STEM education and career opportunities for underrepresented groups, including women and girls. We will work with partners—from the private-sector and nonprofit organizations to universities—to promote education and careers in science and technology.

Increase International Education and Exchange: The pervasiveness of the English language and American cultural influence are great advantages to Americans traveling, working, and negotiating in foreign countries. But we must develop skills to help us succeed in a dynamic and diverse global economy. We will support programs that cultivate interest and scholarship in foreign languages and intercultural affairs, including international exchange programs. This will allow our citizens to build connections with peoples overseas and to develop skills and contacts that will help them thrive in the global economy. We must also welcome more foreign exchange students to our shores, recognizing the benefits that can result from deeper ties with foreign publics and increased understanding of American society.

Pursue Comprehensive Immigration Reform: The United States is a nation of immigrants. Our ability to innovate, our ties to the world, and our economic prosperity depend on our nation's capacity to welcome and assimilate immigrants, and a visa system which welcomes skilled professionals from around the world. At the same time, effective border security and immigration enforcement must keep the country safe and deter unlawful entry. Indeed, persistent problems in immigration policy consume valuable resources needed to advance other security objectives and make it harder to focus on the most dangerous threats facing our country. Ultimately, our national security depends on striking a balance between security and openness. To advance this goal, we must pursue comprehensive immigration reform that effectively secures our borders, while repairing a broken system that fails to serve the needs of our nation.

Enhance Science, Technology, and Innovation

Reaffirming America's role as the global engine of scientific discovery and technological innovation has never been more critical. Challenges like climate change, pandemic disease, and resource scarcity demand new innovation. Meanwhile, the nation that leads the world in building a clean energy economy will enjoy a substantial economic and security advantage. That is why the Administration is investing heavily in research, improving education in science and math, promoting developments in energy, and expanding international cooperation.

Transform our Energy Economy: As long as we are dependent on fossil fuels, we need to ensure the security and free flow of global energy resources. But without significant and timely adjustments, our energy dependence will continue to undermine our security and prosperity. This will leave us vulnerable to energy supply disruptions and manipulation and to changes in the environment on an unprecedented scale.

The United States has a window of opportunity to lead in the development of clean energy technology. If successful, the United States will lead in this new Industrial Revolution in clean energy that will be a major contributor to our economic prosperity. If we do not develop the policies that encourage the private sector to seize the opportunity, the United States will fall behind and increasingly become an importer of these new energy technologies.

We have already made the largest investment in clean energy in history, but there is much more to do to build on this foundation. We must continue to transform our energy economy, leveraging private capital to accelerate deployment of clean energy technologies that will cut greenhouse gas emissions, improve energy efficiency, increase use of renewable and nuclear power, reduce the dependence of vehicles on oil, and diversify energy sources and suppliers. We will invest in research and next-generation technology, modernize the way we distribute electricity, and encourage the usage of transitional fuels, while moving towards clean energy produced at home.

Invest in Research: Research and development is central to our broader national capacity. Incidents like the outbreak of H1N1 influenza and the challenge of identifying new, renewable sources of energy highlight the importance of research in basic and applied science. We are reversing the decades-long decline in federal funding for research, including the single largest infusion to basic science research in American history. Research and innovation is not something government can do on its own, which is why we will support and create incentives to encourage private initiatives. The United States has always excelled in our ability to turn science and technology into engineering and products, and we must continue to do so in the future.

Expand International Science Partnerships: America's scientific leadership has always been widely admired around the world, and we must continue to expand cooperation and partnership in science and technology. We have launched a number of Science Envoys around the globe and are promoting stronger relationships between American scientists, universities, and researchers and their counterparts abroad. We will reestablish a commitment to science and technology in our foreign assistance efforts and develop a strategy for international science and national security.

Employ Technology to Protect our Nation: Our renewed commitment to science and technology—and our ability to apply the ingenuity of our public and private sectors toward the most difficult foreign policy and security challenges of our time—will help us protect our citizens and advance U.S. national security priorities. These include, for example, protecting U.S. and allied forces from asymmetric attacks; supporting arms control and nonproliferation agreements; preventing terrorists from attacking our homeland; preventing and managing widespread disease outbreaks; securing the supply chain; detecting weapons of mass destruction before they reach our borders; and protecting our information, communication, and transportation infrastructure.

Leverage and Grow our Space Capabilities: For over 50 years, our space community has been a catalyst for innovation and a hallmark of U.S. technological leadership. Our space capabilities underpin global commerce and scientific advancements and bolster our national security strengths and those of our allies and partners. To promote security and stability in space, we will pursue activities consistent with the inherent right of self-defense, deepen cooperation with allies and friends, and work with all nations toward the responsible and peaceful use of space. To maintain the advantages afforded to the United States by space, we must also take several actions. We must continue to encourage cutting-edge space technology by investing in the people and industrial base that develops them. We will invest in the research and development of next-generation space technologies and capabilities that benefit our commercial, civil, scientific exploration, and national security communities, in order to maintain the viability of space for future generations. And we will promote a unified effort to strengthen our space industrial base and work with universities to encourage students to pursue space-related careers.

Achieve Balanced and Sustainable Growth

Balanced and sustainable growth, at home and throughout the global economy, drives the momentum of the U.S. economy and underpins our prosperity. A steadily growing global economy means an expanding market for exports of our goods and services. Over time, deepening linkages among markets and businesses will provide the setting in which the energies and entrepreneurship of our private sector can flourish, generating technologies, business growth, and job creation that will boost living standards for Americans. United States economic leadership now has to adapt to the rising prominence of emerging economies; the growing size, speed, and sophistication of financial markets; the multiplicity of market participants around the globe; and the struggling economies that have so far failed to integrate into the global system.

To promote prosperity for all Americans, we will need to lead the international community to expand the inclusive growth of the integrated, global economy. At the same time, we will need to lead international efforts to prevent a recurrence of economic imbalances and financial excesses, while managing the many security threats and global challenges that affect global economic stability. To promote growth that can be balanced and sustained, we will:

Prevent Renewed Instability in the Global Economy: The recent crisis taught us the very high cost of the boom and bust cycle that has plagued the global economy and has served neither the United States nor our international partners. Once Americans found themselves in debt or out of work, our demand for foreign goods fell sharply. As foreign economies weakened, their financial institutions and public finances came under stress too, reinforcing the global slowdown. We must prevent the reemergence of imbalanced growth, with American consumers buying and borrowing, and Asian and other exporting countries selling and accumulating claims. We must pursue reform of the U.S. financial system to strengthen the health of our economy and encourage Americans to save more. And we must prevent the reemergence of excesses in our financial institutions based on irresponsible lending behavior, and abetted by lax and uncoordinated regulation.

Save More and Export More: Striking a better balance at home means saving more and spending less, reforming our financial system, and reducing our long-term budget deficit. With those changes, we will see a greater emphasis on exports that we can build, produce, and sell all over the world, with the goal of doubling U.S. exports by 2014. This is ultimately an employment strategy, because higher exports will support millions of well-paying American jobs, including those that service innovative and profitable new technologies. As a part of that effort, we are reforming our export controls consistent with our national security imperatives.

Shift to Greater Domestic Demand Abroad: For the rest of the world, especially in some emerging market and developing countries, a better balance means placing greater emphasis on increasing domestic demand as the leading driver of growth and opening markets. Those countries will be able to import the capital and technologies needed to sustain the remarkable productivity gains already underway. Rebalancing will provide an opportunity for workers and consumers over time to enjoy the higher standards of living made possible by those gains. As balanced growth translates into sustained growth, middle-income, and poor countries, many of which are not yet sufficiently integrated into the global economy, can accelerate the process of convergence of living standards toward richer countries—a process that will become a driver of growth for the global economy for decades to come.

Open Foreign Markets to Our Products and Services: The United States has long had one of the most open markets in the world. We have been a leader in expanding an open trading system. That has underwritten the growth of other developed and emerging markets alike. Openness has also forced our companies and workers to compete and innovate, and at the same time, has offered market access crucial to the success of so many countries around the world. We will maintain our open investment environment, consistent with our national security goals. In this new era, opening markets around the globe will promote global competition and innovation and will be crucial to our prosperity. We will pursue a trade agenda that includes an ambitious and balanced Doha multilateral trade agreement, bilateral and multilateral trade agreements that reflect our values and interests, and engagement with the transpacific partnership countries to shape a regional agreement with high standards.

As we go forward, our trade policy will be an important part of our effort to capitalize on the opportunities presented by globalization, but will also be part of our effort to equip Americans to compete. To make trade agreements work for Americans, we will take steps to restore confidence, with realistic programs to deal with transition costs, and promote innovation, infrastructure, healthcare reform and education. Our agreements will contain achievable enforcement mechanisms to ensure that the gains we negotiate are in fact realized and will be structured to reflect U.S. interests, especially on labor and environment.

Build Cooperation with Our International Partners: The United States has supported the G-20's emergence as the premier forum for international economic cooperation. This flows from the recognition that we need a broader and more inclusive engagement with the countries responsible for most of global output and trade. U.S. leadership in the G-20 will be focused on securing sustainable and balanced growth, coordinating reform of financial sector regulation, fostering global economic development, and promoting energy security. We also need official international financial institutions to be as modern and agile as the global economy they serve. Through the G-20, we will pursue governance reform at the International Monetary Fund (IMF) and World Bank. We will also broaden our leadership in other international financial institutions so that the rapidly growing countries of the world see their representation increase and are willing to invest those institutions with the authority they need to promote the stability and growth of global output and trade.

Deterring Threats to the International Financial System: Today's open and global financial system also exposes us to global financial threats. Just as we work to make the most of the opportunities that globalization brings, the actors that pose a threat to our national security—terrorists, proliferators, narcotics traffickers, corrupt officials, and others—are abusing the global financial system to raise, move, and safeguard funds that support their illicit activities or from which they derive profit. Their support networks have global reach and are not contained by national borders. Our strategy to attack these networks must respond in kind and target their illicit resources and access to the global financial system through financial measures, administration and enforcement of regulatory authorities, outreach to the private sector and our foreign partners, and collaboration on international standards and information sharing.

Accelerate Sustainable Development

The growth of emerging economies in recent decades has lifted people out of poverty and forged a more interconnected and vibrant global economy. But development has been uneven, progress is fragile, and too many of the world's people still live without the benefits that development affords. While some countries are growing, many lag behind—mired in insecurity, constrained by poor governance, or overly dependent upon commodity prices. But sustained economic progress requires faster, sustainable, and more inclusive development. That is why we are pursuing a range of specific initiatives in areas such as food security and global health that will be essential to the future security and prosperity of nations and peoples around the globe.

Increase Investments in Development: The United States has an interest in working with our allies to help the world's poorest countries grow into productive and prosperous economies governed by capable, democratic, and accountable state institutions. We will ensure a greater and more deliberate focus on a global development agenda across the United States Government, from policy analysis through policy implementation. We are increasing our foreign assistance, expanding our investments in effective multilateral development institutions, and leveraging the engagement of others to share the burden.

Invest in the Foundations of Long-Term Development: The United States will initiate long-term investments that recognize and reward governments that demonstrate the capacity and political will to pursue sustainable development strategies and ensure that all policy instruments at our disposal are harnessed to these ends. And we will provide our support in multiple ways—by strengthening the ability of governments and communities to manage development challenges and investing in strong institutions that foster the democratic accountability that helps sustain development. This will expand the circle of nations—particularly in Africa—who are capable of reaping the benefits of the global economy, while contributing to global security and prosperity.

Exercise Leadership in the Provision of Global Public Goods: Our approach needs to reflect the fact that there are a set of development challenges that strongly affect the likelihood of progress, but cannot be addressed by individual countries acting alone. Particularly in Africa, these challenges—such as adaptation to global warming, the control of epidemic disease, and the knowledge to increase agricultural productivity—are not adequately addressed in bilateral efforts. We will shape the international architecture and work with our global partners to address these challenges, and increase our investments and engagement to transition to a low-carbon growth trajectory, support the resilience of the poorest nations to the effects of climate change, and strengthen food security. We must also pursue potential "game changers" for development such as new vaccines, weather-resistant seed varieties, and green energy technologies.

Spend Taxpayers' Dollars Wisely

The United States Government has an obligation to make the best use of taxpayer money, and our ability to achieve long-term goals depends upon our fiscal responsibility. A responsible budget involves making tough choices to live within our means; holding departments and agencies accountable for their spending and their performance; harnessing technology to improve government performance; and being open and honest with the American people. A responsible budget also depends upon working with our global partners and institutions to share burdens and leverage U.S. investments to achieve global goals. Our national security goals can only be reached if we make hard choices and work with international partners to share burdens.

Reduce the Deficit: We cannot grow our economy in the long term unless we put the United States back on a sustainable fiscal path. To begin this effort, the Administration has proposed a 3-year freeze in nonsecurity discretionary spending, a new fee on the largest financial services companies to recoup taxpayer losses for the Troubled Asset Relief Program (TARP), and the closing of tax loopholes and unnecessary subsidies. The Administration has created a bipartisan fiscal commission to suggest further steps for medium-term deficit reduction and will work for fiscally responsible health insurance reform that will bring down the rate of growth in health care costs, a key driver of the country's fiscal future.

Reform Acquisition and Contracting Processes: Wasteful spending, duplicative programs, and contracts with poor oversight have no place in the United States Government. Cost-effective and efficient processes are particularly important for the Department of Defense, which accounts for approximately 70 percent of all Federal procurement spending. We will scrutinize our programs and terminate or restructure those that are outdated, duplicative, ineffective, or wasteful. The result will be more relevant, capable, and effective programs and systems that our military wants and needs. We are also reforming Federal contracting and strengthening contracting practices and management oversight with a goal of saving Federal agencies $40 billion dollars a year.

Increase Transparency: Americans have a right to know how their tax dollars are spent, but that information can be obscured or unavailable. In some instances, incomplete accounting of the budget has been used to conceal the reality of our fiscal situation. To uphold our commitment to a transparent budget process, we are simultaneously requesting both base budget and overseas contingency operations costs, with the same amount of justification and explanatory material for each, so that Americans can see the true cost of our war efforts and hold leaders accountable for decisions with all of the facts.

Values

> "We uphold our most cherished values not only because doing so is right, but because it strengthens our country and keeps us safe. Time and again, our values have been our best national security asset—in war and peace, in times of ease, and in eras of upheaval. Fidelity to our values is the reason why the United States of America grew from a small string of colonies under the writ of an empire to the strongest nation in the world."
>
> —President Barack Obama, National Archives, May 21, 2009

The United States believes certain values are universal and will work to promote them worldwide. These include an individual's freedom to speak their mind, assemble without fear, worship as they please, and choose their own leaders; they also include dignity, tolerance, and equality among all people, and the fair and equitable administration of justice. The United States was founded upon a belief in these values. At home, fidelity to these values has extended the promise of America ever more fully, to ever more people. Abroad, these values have been claimed by people of every race, region, and religion. Most nations are parties to international agreements that recognize this commonality. And nations that embrace these values for their citizens are ultimately more successful—and friendly to the United States—than those that do not.

Yet after an era that saw substantial gains for these values around the world, democratic development has stalled in recent years. In some cultures, these values are being equated with the ugly face of modernity and are seen to encroach upon cherished identities. In other countries, autocratic rulers have repressed basic human rights and democratic practices in the name of economic development and national unity. Even where some governments have adopted democratic practices, authoritarian rulers have undermined electoral processes and restricted the space for opposition and civil society, imposing a growing number of legal restrictions so as to impede the rights of people to assemble and to access information. And while there has been substantial progress in combating poverty in many parts of the world, too many of the world's people still lack the dignity that comes with the opportunity to pursue a better life.

The United States supports those who seek to exercise universal rights around the world. We promote our values above all by living them at home. We continue to engage nations, institutions, and peoples in pursuit of these values abroad. And we recognize the link between development and political progress. In doing so, our goals are realistic, as we recognize that different cultures and traditions give life to these values in distinct ways. Moreover, America's influence comes not from perfection, but from our striving to overcome our imperfections. The constant struggle to perfect our union is what makes the American story inspiring. That is why acknowledging our past shortcomings —and highlighting our efforts to remedy them—is a means of promoting our values.

America will not impose any system of government on another country, but our long-term security and prosperity depends on our steady support for universal values, which sets us apart from our enemies, adversarial governments, and many potential competitors for influence. We will do so through a variety of means—by speaking out for universal rights, supporting fragile democracies and civil society, and supporting the dignity that comes with development.

Strengthen the Power of Our Example

More than any other action that we have taken, the power of America's example has helped spread freedom and democracy abroad. That is why we must always seek to uphold these values not just when it is easy, but when it is hard. Advancing our interests may involve new arrangements to confront threats like terrorism, but these practices and structures must always be in line with our Constitution, preserve our people's privacy and civil liberties, and withstand the checks and balances that have served us so well. To sustain our fidelity to our values—and our credibility to promote them around the world—we will continue to:

Prohibit Torture without Exception or Equivocation: Brutal methods of interrogation are inconsistent with our values, undermine the rule of law, and are not effective means of obtaining information. They alienate the United States from the world. They serve as a recruitment and propaganda tool for terrorists. They increase the will of our enemies to fight against us, and endanger our troops when they are captured. The United States will not use or support these methods.

Legal Aspects of Countering Terrorism: The increased risk of terrorism necessitates a capacity to detain and interrogate suspected violent extremists, but that framework must align with our laws to be effective and sustainable. When we are able, we will prosecute terrorists in Federal courts or in reformed military commissions that are fair, legitimate, and effective. For detainees who cannot be prosecuted—but pose a danger to the American people—we must have clear, defensible, and lawful standards. We must have fair procedures and a thorough process of periodic review, so that any prolonged detention is carefully evaluated and justified. And keeping with our Constitutional system, it will be subject to checks and balances. The goal is an approach that can be sustained by future Administrations, with support from both political parties and all three branches of government.

Balance the Imperatives of Secrecy and Transparency: For the sake of our security, some information must be protected from public disclosure—for instance, to protect our troops, our sources and methods of intelligence-gathering or confidential actions that keep the American people safe. Yet our democracy depends upon transparency, and whenever possible, we are making information available to the American people so that they can make informed judgments and hold their leaders accountable. For instance, when we invoke the State Secrets privilege, we will follow clear procedures so as to provide greater accountability and to ensure the privilege is invoked only when necessary and in the narrowest way possible. We will never invoke the privilege to hide a violation of law or to avoid embarrassment to the government.

Protect Civil Liberties, Privacy, and Oversight: Protecting civil liberties and privacy are integral to the vibrancy of our democracy and the exercise of freedom. We are balancing our solemn commitments to these virtues with the mandate to provide security for the American people. Vigorous oversight of national security activities by our three branches of government and vigilant compliance with the rule of law allow us to maintain this balance, affirm to our friends and allies the constitutional ideals we uphold.

Uphold the Rule of Law: The rule of law—and our capacity to enforce it—advances our national security and strengthens our leadership. At home, fidelity to our laws and support for our law enforcement community safeguards American citizens and interests, while protecting and advancing our values. Around the globe, it allows us to hold actors accountable, while supporting both international security and the stability of the global economy. America's commitment to the rule of law is fundamental to our efforts to build an international order that is capable of confronting the emerging challenges of the 21st century.

Draw Strength from Diversity: The United States has benefited throughout our history when we have drawn strength from our diversity. While those who advocate on behalf of extremist ideologies seek to sow discord among ethnic and religious groups, America stands as an example of how people from different backgrounds can be united through their commitment to shared values. Within our own communities, those who seek to recruit and radicalize individuals will often try to prey upon isolation and alienation. Our own commitment to extending the promise of America will both draw a contrast with those who try to drive people apart, while countering attempts to enlist individuals in ideological, religious, or ethnic extremism.

Promote Democracy and Human Rights Abroad

The United States supports the expansion of democracy and human rights abroad because governments that respect these values are more just, peaceful, and legitimate. We also do so because their success abroad fosters an environment that supports America's national interests. Political systems that protect universal rights are ultimately more stable, successful, and secure. As our history shows, the United States can more effectively forge consensus to tackle shared challenges when working with governments that reflect the will and respect the rights of their people, rather than just the narrow interests of those in power. The United States is advancing universal values by:

Ensuring that New and Fragile Democracies Deliver Tangible Improvements for Their Citizens: The United States must support democracy, human rights, and development together, as they are mutually reinforcing. We are working closely with citizens, communities, and political and civil society leaders to strengthen key institutions of democratic accountability—free and fair electoral processes, strong legislatures, civilian control of militaries, honest police forces, independent and fair judiciaries, a free and independent press, a vibrant private sector, and a robust civil society. To do so, we are harnessing our bilateral and multilateral capabilities to help nascent democracies deliver services that respond to the needs and preferences of their citizens, since democracies without development rarely survive.

Practicing Principled Engagement with Non-Democratic Regimes: Even when we are focused on interests such as counterterrorism, nonproliferation, or enhancing economic ties, we will always seek in parallel to expand individual rights and opportunities through our bilateral engagement. The United States is pursuing a dual-track approach in which we seek to improve government-to-government relations and use this dialogue to advance human rights, while engaging civil society and peaceful political opposition, and encouraging U.S. nongovernmental actors to do the same. More substantive government-to-government relations can create permissive conditions for civil society to operate and for more extensive people-to-people exchanges. But when our overtures are rebuffed, we must lead the international community in using public and private diplomacy, and drawing on incentives and disincentives, in an effort to change repressive behavior.

Recognizing the Legitimacy of All Peaceful Democratic Movements: America respects the right of all peaceful, law-abiding, and nonviolent voices to be heard around the world, even if we disagree with them. Support for democracy must not be about support for specific candidates or movements. America will welcome all legitimately elected, peaceful governments, provided they govern with respect for the rights and dignity of all their people and consistent with their international obligations. Those who seek democracy to obtain power, but are ruthless once they do, will forfeit the support of the United States. Governments must maintain power through consent, not coercion, and place legitimate political processes above party or narrow interest.

Supporting the Rights of Women and Girls: Women should have access to the same opportunities and be able to make the same choices as men. Experience shows that countries are more peaceful and prosperous when women are accorded full and equal rights and opportunity. When those rights and opportunities are denied, countries often lag behind. Furthermore, women and girls often disproportionally bear the burden of crises and conflict. Therefore the United States is working with regional and international organizations to prevent violence against women and girls, especially in conflict zones. We are supporting women's equal access to justice and their participation in the political process. We are promoting child and maternal health. We are combating human trafficking, especially in women and girls, through domestic and international law enforcement. And we are supporting education, employment, and micro-finance to empower women globally.

Strengthening International Norms Against Corruption: We are working within the broader international system, including the U.N., G-20, Organization for Economic Cooperation and Development (OECD), and the international financial institutions, to promote the recognition that pervasive corruption is a violation of basic human rights and a severe impediment to development and global security. We will work with governments and civil society organizations to bring greater transparency and accountability to government budgets, expenditures, and the assets of public officials. And we will institutionalize transparent practices in international aid flows, international banking and tax policy, and private sector engagement around natural resources to make it harder for officials to steal and to strengthen the efforts of citizens to hold their governments accountable.

Building a Broader Coalition of Actors to Advance Universal Values: We are working to build support for democracy, rule of law, and human rights by working with other governments, nongovernmental organizations, and multilateral fora. The United States is committed to working to shape and strengthen existing institutions that are not delivering on their potential, such as the United Nations Human Rights Council. We are working within the broader U.N. system and through regional mechanisms to strengthen human rights monitoring and enforcement mechanisms, so that individuals and countries are held accountable for their violation of international human rights norms. And we will actively support the leadership of emerging democracies as they assume a more active role in advancing basic human rights and democratic values in their regions and on the global stage.

Marshalling New Technologies and Promoting the Right to Access Information: The emergence of technologies such as the Internet, wireless networks, mobile smart-phones, investigative forensics, satellite and aerial imagery, and distributed remote sensing infrastructure has created powerful new opportunities to advance democracy and human rights. These technologies have fueled people-powered political movements, made it possible to shine a spotlight on human rights abuses nearly instantaneously, and increased avenues for free speech and unrestricted communication around the world. We support the dissemination and use of these technologies to facilitate freedom of expression, expand access to information, increase governmental transparency and accountability, and counter restrictions on their use. We will also better utilize such technologies to effectively communicate our own messages to the world.

Promote Dignity by Meeting Basic Needs

The freedom that America stands for includes freedom from want. Basic human rights cannot thrive in places where human beings do not have access to enough food, or clean water, or the medicine they need to survive. The United States has embraced the United Nation's Millennium Development Goals and is working with others in pursuit of the eradication of extreme poverty—efforts that are particularly critical to the future of nations and peoples of Africa. And we will continue to promote the dignity that comes through development efforts such as:

Pursuing a Comprehensive Global Health Strategy: The United States has a moral and strategic interest in promoting global health. When a child dies of a preventable disease, it offends our conscience; when a disease goes unchecked, it can endanger our own health; when children are sick, development is stalled. That is why we are continuing to invest in the fight against HIV/AIDS. Through the Global Health Initiative, we will strengthen health systems and invest in interventions to address areas where progress has lagged, including maternal and child health. And we are also pursuing the goal of reducing the burden of malaria and tuberculosis and seeking the elimination of important neglected tropical diseases.

Promoting Food Security: The United States is working with partners around the world to advance a food security initiative that combats hunger and builds the capacity of countries to feed their people. Instead of simply providing aid for developing countries, we are focusing on new methods and technologies for agricultural development. This is consistent with an approach in which aid is not an end in itself—the purpose of our foreign assistance will be to create the conditions where it is no longer needed.

Leading Efforts to Address Humanitarian Crises: Together with the American people and the international community, we will continue to respond to humanitarian crises to ensure that those in need have the protection and assistance they need. In such circumstances, we are also placing a greater emphasis on fostering long-term recovery. Haiti's devastating earthquake is only the most recent reminder of the human and material consequences of natural disasters, and a changing climate portends a future in which the United States must be better prepared and resourced to exercise robust leadership to help meet critical humanitarian needs.

International Order

> "As President of the United States, I will work tirelessly to protect America's security and to advance our interests. But no one nation can meet the challenges of the 21st century on its own, nor dictate its terms to the world. That is why America seeks an international system that lets nations pursue their interests peacefully, especially when those interests diverge; a system where the universal rights of human beings are respected, and violations of those rights are opposed; a system where we hold ourselves to the same standards that we apply to other nations, with clear rights and responsibilities for all."
>
> —President Barack Obama, Moscow, Russia, July 7, 2009

The United States will protect its people and advance our prosperity irrespective of the actions of any other nation, but we have an interest in a just and sustainable international order that can foster collective action to confront common challenges. This international order will support our efforts to advance security, prosperity, and universal values, but it is also an end that we seek in its own right. Because without such an international order, the forces of instability and disorder will undermine global security. And without effective mechanisms to forge international cooperation, challenges that recognize no borders—such as climate change, pandemic disease, and transnational crime—will persist and potentially spread.

International institutions—most prominently NATO and the United Nations—have been at the center of our international order since the mid 20th century. Yet, an international architecture that was largely forged in the wake of World War II is buckling under the weight of new threats, making us less able to seize new opportunities. Even though many defining trends of the 21st century affect all nations and peoples, too often, the mutual interests of nations and peoples are ignored in favor of suspicion and self-defeating competition.

What is needed, therefore, is a realignment of national actions and international institutions with shared interests. And when national interests do collide—or countries prioritize their interests in different ways—those nations that defy international norms or fail to meet their sovereign responsibilities will be denied the incentives that come with greater integration and collaboration with the international community.

No international order can be supported by international institutions alone. Our mutual interests must be underpinned by bilateral, multilateral, and global strategies that address underlying sources of insecurity and build new spheres of cooperation. To that end, strengthening bilateral and multilateral cooperation cannot be accomplished simply by working inside formal institutions and frameworks. It requires sustained outreach to foreign governments, political leaderships, and other critical constituencies that must commit the necessary capabilities and resources to enable effective, collective action. And it means building upon our traditional alliances, while also cultivating partnerships with new centers of influence. Taken together, these approaches will allow us to foster more effective global cooperation to confront challenges that know no borders and affect every nation.

Ensure Strong Alliances

The foundation of United States, regional, and global security will remain America's relations with our allies, and our commitment to their security is unshakable. These relationships must be constantly cultivated, not just because they are indispensible for U.S. interests and national security objectives, but because they are fundamental to our collective security. Alliances are force multipliers: through multinational cooperation and coordination, the sum of our actions is always greater than if we act alone. We will continue to maintain the capacity to defend our allies against old and new threats. We will also continue to closely consult with our allies as well as newly emerging partners and organizations so that we revitalize and expand our cooperation to achieve common objectives. And we will continue to mutually benefit from the collective security provided by strong alliances.

Although the United States and our allies and partners may sometimes disagree on specific issues, we will act based upon mutual respect and in a manner that continues to strengthen an international order that benefits all responsible international actors.

Strengthening Security Relationships: Our ability to sustain these alliances, and to build coalitions of support toward common objectives, depends in part on the capabilities of America's Armed Forces. Similarly, the relationships our Armed Forces have developed with foreign militaries are a critical component of our global engagement and support our collective security.

We will continue to ensure that we can prevail against a wide range of potential adversaries—to include hostile states and nonstate actors—while broadly shaping the strategic environment using all tools to advance our common security. We will continue to reassure our allies and partners by retaining our ability to bring precise, sustained, and effective capabilities to bear against a wide range of military threats and decisively defeat the forces of hostile regional powers. We will work with our allies and partners to enhance the resilience of U.S. forward posture and facilities against potential attacks. Finally, we will strengthen our regional deterrence postures—for example, through phased, adaptive missile defense architectures—in order to make certain that regional adversaries gain no advantages from their acquisition of new, offensive military capabilities.

European Allies: Our relationship with our European allies remains the cornerstone for U.S. engagement with the world, and a catalyst for international action. We will engage with our allies bilaterally, and pursue close consultation on a broad range of security and economic issues. The North Atlantic Treaty Organization (NATO) is the pre-eminent security alliance in the world today. With our 27 NATO allies, and the many partners with which NATO cooperates, we will strengthen our collective ability to promote security, deter vital threats, and defend our people. NATO's new Strategic Concept will provide an opportunity to revitalize and reform the Alliance. We are committed to ensuring that NATO is able to address the full range of 21st century challenges, while serving as a foundation of European security. And we will continue to anchor our commitment in Article V, which is fundamental to our collective security.

Building on European aspirations for greater integration, we are committed to partnering with a stronger European Union to advance our shared goals, especially in promoting democracy and prosperity in Eastern European countries that are still completing their democratic transition and in responding to pressing issues of mutual concern. We will remain dedicated to advancing stability and democracy in the Balkans and to resolving conflicts in the Caucasus and in Cyprus. We will continue to engage with Turkey on a broad range of mutual goals, especially with regard to pursuit of stability in its region. And we will seek to strengthen existing European institutions so that they are more inclusive and more effective in building confidence, reducing tensions, and protecting freedom.

Asian Allies: Our alliances with Japan, South Korea, Australia, the Philippines, and Thailand are the bedrock of security in Asia and a foundation of prosperity in the Asia-Pacific region. We will continue to deepen and update these alliances to reflect the dynamism of the region and strategic trends of the 21st century. Japan and South Korea are increasingly important leaders in addressing regional and global issues, as well as in embodying and promoting our common democratic values. We are modernizing our security relationships with both countries to face evolving 21st century global security challenges and to reflect the principle of equal partnership with the United States and to ensure a sustainable foundation for the U.S. military presence there. We are working together with our allies to develop a positive security agenda for the region, focused on regional security, combating the proliferation of weapons of mass destruction, terrorism, climate change, international piracy, epidemics, and cybersecurity, while achieving balanced growth and human rights.

In partnership with our allies, the United States is helping to offer a future of security and integration to all Asian nations and to uphold and extend fundamental rights and dignity to all of its people. These alliances have preserved a hard-earned peace and strengthened the bridges of understanding across the Pacific Ocean in the second half of the 20th century, and it is essential to U.S., Asian, and global security that they are as dynamic and effective in the 21st century.

North America: The strategic partnerships and unique relationships we maintain with Canada and Mexico are critical to U.S. national security and have a direct effect on the security of our homeland. With billions of dollars in trade, shared critical infrastructure, and millions of our citizens moving across our common borders, no two countries are more directly connected to our daily lives. We must change the way we think about our shared borders, in order to secure and expedite the lawful and legitimate flow of people and goods while interdicting transnational threat that threaten our open societies.

Canada is our closest trading partner, a steadfast security ally, and an important partner in regional and global efforts. Our mutual prosperity is closely interconnected, including through our trade relationship with Mexico through NAFTA. With Canada, our security cooperation includes our defense of North America and our efforts through NATO overseas. And our cooperation is critical to the success of international efforts on issues ranging from international climate negotiations to economic cooperation through the G-20.

With Mexico, in addition to trade cooperation, we are working together to identify and interdict threats at the earliest opportunity, even before they reach North America. Stability and security in Mexico are indispensable to building a strong economic partnership, fighting the illicit drug and arms trade, and promoting sound immigration policy.

Build Cooperation with Other 21st Century Centers of Influence

The United States is part of a dynamic international environment, in which different nations are exerting greater influence, and advancing our interests will require expanding spheres of cooperation around the word. Certain bilateral relationships—such as U.S. relations with China, India, and Russia—will be critical to building broader cooperation on areas of mutual interest. And emerging powers in every region of the world are increasingly asserting themselves, raising opportunities for partnership for the United States.

Asia: Asia's dramatic economic growth has increased its connection to America's future prosperity, and its emerging centers of influence make it increasingly important. We have taken substantial steps to deepen our engagement in the region, through regional organizations, new dialogues, and high-level diplomacy. The United States has deep and enduring ties with the countries of the region, including trade and investment that drive growth and prosperity on both sides of the Pacific, and enhancing these ties is critical to our efforts to advance balanced and sustainable growth and to doubling U.S. exports. We have increasing security cooperation on issues such as violent extremism and nuclear proliferation. We will work to advance these mutual interests through our alliances, deepen our relationships with emerging powers, and pursue a stronger role in the region's multilateral architecture, including the Association of Southeast Asian Nations (ASEAN), the Asia Pacific Economic Cooperation forum, the Trans-Pacific Partnership, and the East Asia Summit.

We will continue to pursue a positive, constructive, and comprehensive relationship with China. We welcome a China that takes on a responsible leadership role in working with the United States and the international community to advance priorities like economic recovery, confronting climate change, and nonproliferation. We will monitor China's military modernization program and prepare accordingly to ensure that U.S. interests and allies, regionally and globally, are not negatively affected. More broadly, we will encourage China to make choices that contribute to peace, security, and prosperity as its influence rises. We are using our newly established Strategic and Economic Dialogue to address a broader range of issues, and improve communication between our militaries in order to reduce mistrust. We will encourage continued reduction in tension between the People's Republic of China and Taiwan. We will not agree on every issue, and we will be candid on our human rights concerns and areas where we differ. But disagreements should not prevent cooperation on issues of mutual interest, because a pragmatic and effective relationship between the United States and China is essential to address the major challenges of the 21st century.

The United States and India are building a strategic partnership that is underpinned by our shared interests, our shared values as the world's two largest democracies, and close connections among our people. India's responsible advancement serves as a positive example for developing nations, and provides an opportunity for increased economic, scientific, environmental, and security partnership. Working together through our Strategic Dialogue and high-level visits, we seek a broad-based relationship in which India contributes to global counterterrorism efforts, nonproliferation, and helps promote poverty-reduction, education, health, and sustainable agriculture. We value India's growing leadership on a wide array of global issues, through groups such as the G-20, and will seek to work with India to promote stability in South Asia and elsewhere in the world.

Russia: We seek to build a stable, substantive, multidimensional relationship with Russia, based on mutual interests. The United States has an interest in a strong, peaceful, and prosperous Russia that respects international norms. As the two nations possessing the majority of the world's nuclear weapons, we are working together to advance nonproliferation, both by reducing our nuclear arsenals and by cooperating to ensure that other countries meet their international commitments to reducing the spread of nuclear weapons around the world. We will seek greater partnership with Russia in confronting violent extremism, especially in Afghanistan. We also will seek new trade and investment arrangements for increasing the prosperity of our peoples. We support efforts within Russia to promote the rule of law, accountable government, and universal values. While actively seeking Russia's cooperation to act as a responsible partner in Europe and Asia, we will support the sovereignty and territorial integrity of Russia's neighbors.

Emerging Centers of Influence: Due to increased economic growth and political stability, individual nations are increasingly taking on powerful regional and global roles and changing the landscape of international cooperation. To achieve a just and sustainable order that advances our shared security and prosperity, we are, therefore, deepening our partnerships with emerging powers and encouraging them to play a greater role in strengthening international norms and advancing shared interests.

The rise of the G-20, for example, as the premier international economic forum, represents a distinct shift in our global international order toward greater cooperation between traditional major economies and emerging centers of influence. The nations composing the G-20—from South Korea to South Africa, Saudi Arabia to Argentina—represent at least 80 percent of global gross national product, making it an influential body on the world stage. Stabilizing our global economy, increasing energy efficiency around the globe, and addressing chronic hunger in poor countries are only three examples of the broad global challenges that cannot be solved by a few countries alone.

Indonesia—as the world's fourth most populous country, a member of the G-20, and a democracy—will become an increasingly important partner on regional and transnational issues such as climate change, counterterrorism, maritime security, peacekeeping, and disaster relief. With tolerance, resilience, and multiculturalism as core values, and a flourishing civil society, Indonesia is uniquely positioned to help address challenges facing the developing world.

In the Americas, we are bound by proximity, integrated markets, energy interdependence, a broadly shared commitment to democracy, and the rule of law. Our deep historical, familial, and cultural ties make our alliances and partnerships critical to U.S. interests. We will work in equal partnership to advance economic and social inclusion, safeguard citizen safety and security, promote clean energy, and defend universal values of the people of the hemisphere.

We welcome Brazil's leadership and seek to move beyond dated North-South divisions to pursue progress on bilateral, hemispheric, and global issues. Brazil's macroeconomic success, coupled with its steps to narrow socioeconomic gaps, provide important lessons for countries throughout the Americas and Africa. We will encourage Brazilian efforts against illicit transnational networks. As guardian of a unique national environmental patrimony and a leader in renewable fuels, Brazil is an important partner in confronting global climate change and promoting energy security. And in the context of the G-20 and the Doha round, we will work with Brazil to ensure that economic development and prosperity is broadly shared.

We have an array of enduring interests, longstanding commitments and new opportunities for broadening and deepening relationships in the greater Middle East. This includes maintaining a strong partnership with Israel while supporting Israel's lasting integration into the region. The U.S. also will continue to develop our key security relationships in the region with such Arab states as with Egypt, Jordan, and Saudi Arabia and other Gulf Cooperation Council (GCC) countries—partnerships that enable our militaries and defense systems to work together more effectively.

We have a strategic interest in ensuring that the social and economic needs and political rights of people in this region, who represent one of the world's youngest populations, are met. We will continue to press governments in the region to undertake political reforms and to loosen restrictions on speech, assembly and media. We will maintain our strong support for civil society groups and those individuals who stand up for universal rights. And we will continue to foster partnerships in areas like education, economic growth, science, and health to help expand opportunity. On a multilateral basis, we seek to advance shared security interests, such as through NATO's Istanbul Cooperation Initiative with the GCC, and common interests in promoting governance and institutional reform through participating in the Forum for the Future and other regional dialogues.

The diversity and complexity of the African continent offer the United States opportunities and challenges. As African states grow their economies and strengthen their democratic institutions and governance, America will continue to embrace effective partnerships. Our economic, security, and political cooperation will be consultative and encompass global, regional, and national priorities including access to open markets, conflict prevention, global peacekeeping, counterterrorism, and the protection of vital carbon sinks. The Administration will refocus its priorities on strategic interventions that can promote job creation and economic growth; combat corruption while strengthening good governance and accountability; responsibly improve the capacity of African security and rule of law sectors; and work through diplomatic dialogue to mitigate local and regional tensions before they become crises. We will also reinforce sustainable stability in key states like Nigeria and Kenya that are essential subregional linchpins.

The United States will work to remain an attractive and influential partner by ensuring that African priorities such as infrastructure development, improving reliable access to power, and increased trade and investment remain high on our agenda. South Africa's inclusion in the G-20 should be followed by a growing number of emerging African nations who are charting a course toward improved governance and meaningful development. South Africa's vibrant democracy, combined with its regional and global leadership roles, is a critical partner. From peacemaking to climate change to capacity-building, South Africa brings unique value and perspective to international initiatives. With its strong, diversified, well- managed economy, it often serves as a springboard to the entire African continent, and we will work to pursue shared interests in Africa's security, growth, and the development of Africa's human capital.

Strengthen Institutions and Mechanisms for Cooperation

Just as U.S. foresight and leadership were essential to forging the architecture for international cooperation after World War II, we must again lead global efforts to modernize the infrastructure for international cooperation in the 21st century. Indeed, our ability to advance peace, security, and opportunity will turn on our ability to strengthen both our national and our multilateral capabilities. To solve problems, we will pursue modes of cooperation that reflect evolving distributions of power and responsibility. We need to assist existing institutions to perform effectively. When they come up short, we must seek meaningful changes and develop alternative mechanisms.

Enhance Cooperation with and Strengthen the United Nations: We are enhancing our coordination with the U.N. and its agencies. We need a U.N. capable of fulfilling its founding purpose—maintaining international peace and security, promoting global cooperation, and advancing human rights. To this end, we are paying our bills. We are intensifying efforts with partners on and outside the U.N. Security Council to ensure timely, robust, and credible Council action to address threats to peace and security. We favor Security Council reform that enhances the U.N.'s overall performance, credibility, and legitimacy. Across the broader U.N. system we support reforms that promote effective and efficient leadership and management of the U.N.'s international civil service, and we are working with U.N. personnel and member states to strengthen the U.N.'s leadership and operational capacity in peacekeeping, humanitarian relief, post-disaster recovery, development assistance, and the promotion of human rights. And we are supporting new U.N. frameworks and capacities for combating transnational threats like proliferation of weapons of mass destruction, infectious disease, drug-trafficking, and counterterrorism.

Pursue Decisions though a Wide Range of Frameworks and Coalitions: We need to spur and harness a new diversity of instruments, alliances, and institutions in which a division of labor emerges on the basis of effectiveness, competency, and long-term reliability. This requires enhanced coordination among the United Nations, regional organizations, international financial institutions, specialized agencies, and other actors that are better placed or equipped to manage certain threats and challenges. We are attempting to forge new agreement on common global challenges among the world's leading and emerging powers to ensure that multilateral cooperation reflects the sustained commitment of infl uential countries. While we are pursuing G-8 initiatives with proven and long-standing partners, have begun to shift the focus of our economic coordination to the G-20, which is more reflective of today's diffusion of power and the need to enlist the efforts of a broader spectrum of countries across Asia to Europe, Africa to the Middle East, and our neighbors in the Americas. We are also renewing U.S. leadership in the multilateral development banks and the IMF, and leveraging our engagement and investments in these institutions to strengthen the global economy, lift people out of poverty, advance food security, address climate and pandemics, and secure fragile states such as Afghanistan and Haiti.

Invest in Regional Capabilities: Regional organizations can be particularly effective at mobilizing and legitimating cooperation among countries closest to the problem. Regional organizations—whether NATO, the Organization for Security Cooperation in Europe, the Organization of the Islamic Conference, the African Union, Organization of American States, or ASEAN, and the Gulf Cooperation Council—vary widely in their membership, constitutions, histories, orientation, and operational capabilities. That variety needs to inform a strategic approach to their evolving roles and relative contributions to global security.

The United States is encouraging continued innovation and development of enhanced regional capabilities in the context of an evolving division of labor among local, national, and global institutions that seeks to leverage relative capacities. Where appropriate, we use training and related programs to strengthen regional capacities for peacekeeping and conflict management to improve impact and share burdens. We will also encourage a more comprehensive approach to regional security that brings balanced focus to issues such as food security, global health, and education; access to more affordable and greener forms of energy; access to fair and efficient justice; and a concerted effort to promote transparency at all levels and to fight the corrosive effect of corruption.

Sustain Broad Cooperation on Key Global Challenges

Many of today's challenges cannot be solved by one nation or even a group of nations. The test of our international order, therefore, will be its ability to facilitate the broad and effective global cooperation necessary to meet 21st century challenges. Many of these challenges have been discussed previously, including violent extremism, nuclear proliferation, and promotion of global prosperity. In addition, other key challenges requiring broad global cooperation include:

Climate Change: The danger from climate change is real, urgent, and severe. The change wrought by a warming planet will lead to new conflicts over refugees and resources; new suffering from drought and famine; catastrophic natural disasters; and the degradation of land across the globe. The United States will therefore confront climate change based upon clear guidance from the science, and in cooperation with all nations—for there is no effective solution to climate change that does not depend upon all nations taking responsibility for their own actions and for the planet we will leave behind.

- **Home:** Our effort begins with the steps that we are taking at home. We will stimulate our energy economy at home, reinvigorate the U.S. domestic nuclear industry, increase our efficiency standards, invest in renewable energy, and provide the incentives that make clean energy the profitable kind of energy. This will allow us to make deep cuts in emissions—in the range of 17 percent by 2020 and more than 80 percent by 2050. This will depend in part upon comprehensive legislation and its effective implementation.
- **Abroad:** Regionally, we will build on efforts in Asia, the Americas, and Africa to forge new clean energy partnerships. Globally, we will seek to implement and build on the Copenhagen Accord, and ensure a response to climate change that draws upon decisive action by all nations. Our goal is an effective, international effort in which all major economies commit to ambitious national action to reduce their emissions, nations meet their commitments in a transparent manner, and the necessary financing is mobilized so that developing countries can adapt to climate change, mitigate its impacts, conserve forests, and invest in clean energy technologies. We will pursue this global cooperation through multiple avenues, with a focus on advancing cooperation that works. We accept the principle of common but differentiated responses and respective capabilities, but will insist that any approach draws upon each nation taking responsibility for its own actions.

Peacekeeping and Armed Conflict: The untold loss of human life, suffering, and property damage that results from armed conflict necessitates that all responsible nations work to prevent it. No single nation can or should shoulder the burden for managing or resolving the world's armed conflicts. To this end, we will place renewed emphasis on deterrence and prevention by mobilizing diplomatic action, and use development and security sector assistance to build the capacity of at-risk nations and reduce the appeal of violent extremism. But when international forces are needed to respond to threats and keep the peace, we will work with international partners to ensure they are ready, able, and willing. We will continue to build support in other countries to contribute to sustaining global peace and stability operations, through U.N. peacekeeping and regional organizations, such as NATO and the African Union. We will continue to broaden the pool of troop and police contributors, working to ensure that they are properly trained and equipped, that their mandates are matched to means, and that their missions are backed by the political action necessary to build and sustain peace.

In Sudan, which has been marred by violent conflict for decades, the United States remains committed to working with the international community to support implementation of outstanding elements of the Comprehensive Peace Agreement and ensure that the referendum on the future of Southern Sudan in 2011 happens on time and that its results are respected. In addition, we will continue to engage in the efforts necessary to support peace and stability after the referendum, and continue to work to secure peace, dignity, and accountability in Darfur.

- **Prevent Genocide and Mass Atrocities:** The United States and all member states of the U.N. have endorsed the concept of the "Responsibility to Protect." In so doing, we have recognized that the primary responsibility for preventing genocide and mass atrocity rests with sovereign governments, but that this responsibility passes to the broader international community when sovereign governments themselves commit genocide or mass atrocities, or when they prove unable or unwilling to take necessary action to prevent or respond to such crimes inside their borders. The United States is committed to working with our allies, and to strengthening our own internal capabilities, in order to ensure that the United States and the international community are proactively engaged in a strategic effort to prevent mass atrocities and genocide. In the event that prevention fails, the United States will work both multilaterally and bilaterally to mobilize diplomatic, humanitarian, financial, and—in certain instances—military means to prevent and respond to genocide and mass atrocities.
- **International Justice:** From Nuremberg to Yugoslavia to Liberia, the United States has seen that the end of impunity and the promotion of justice are not just moral imperatives; they are stabilizing forces in international affairs. The United States is thus working to strengthen national justice systems and is maintaining our support for ad hoc international tribunals and hybrid courts. Those who intentionally target innocent civilians must be held accountable, and we will continue to support institutions and prosecutions that advance this important interest. Although the United States is not at present a party to the Rome Statute of the International Criminal Court (ICC), and will always protect U.S. personnel, we are engaging with State Parties

to the Rome Statute on issues of concern and are supporting the ICC's prosecution of those cases that advance U.S. interests and values, consistent with the requirements of U.S. law.

Pandemics and Infectious Disease: The threat of contagious disease transcends political boundaries, and the ability to prevent, quickly detect and contain outbreaks with pandemic potential has never been so important. An epidemic that begins in a single community can quickly evolve into a multinational health crisis that causes millions to suffer, as well as spark major disruptions to travel and trade. Addressing these transnational risks requires advance preparation, extensive collaboration with the global community, and the development of a resilient population at home.

Recognizing that the health of the world's population has never been more interdependent, we are improving our public health and medical capabilities on the front lines, including domestic and international disease surveillance, situational awareness, rapid and reliable development of medical countermeasures to respond to public health threats, preparedness education and training, and surge capacity of the domestic health care system to respond to an influx of patients due to a disaster or emergency. These capabilities include our ability to work with international partners to mitigate and contain disease when necessary.

We are enhancing international collaboration and strengthening multilateral institutions in order to improve global surveillance and early warning capabilities and quickly enact control and containment measures against the next pandemic threat. We continue to improve our understanding of emerging diseases and help develop environments that are less conducive to epidemic emergence. We depend on U.S. overseas laboratories, relationships with host nation governments, and the willingness of states to share health data with nongovernmental and international organizations. In this regard, we need to continue to work to overcome the lack of openness and a general reluctance to share health information. Finally, we seek to mitigate other problem areas, including limited global vaccine production capacity, and the threat of emergent and reemergent disease in poorly governed states.

Transnational Criminal Threats and Threats to Governance: Transnational criminal threats and illicit trafficking networks continue to expand dramatically in size, scope, and influence—posing significant national security challenges for the United States and our partner countries. These threats cross borders and continents and undermine the stability of nations, subverting government institutions through corruption and harming citizens worldwide. Transnational criminal organizations have accumulated unprecedented wealth and power through trafficking and other illicit activities, penetrating legitimate financial systems and destabilizing commercial markets. They extend their reach by forming alliances with government officials and some state security services. The crime-terror nexus is a serious concern as terrorists use criminal networks for logistical support and funding. Increasingly, these networks are involved in cyber crime, which cost consumers billions of dollars annually, while undermining global confidence in the international financial system.

Combating transnational criminal and trafficking networks requires a multidimensional strategy that safeguards citizens, breaks the financial strength of criminal and terrorist networks, disrupts illicit trafficking networks, defeats transnational criminal organizations, fights government corruption, strengthens the rule of law, bolsters judicial systems, and improves transparency. While these are major challenges, the United States will be able to devise and execute a collective strategy with other nations facing the same threats.

Safeguarding the Global Commons: Across the globe, we must work in concert with allies and partners to optimize the use of shared sea, air, and space domains. These shared areas, which exist outside exclusive national jurisdictions, are the connective tissue around our globe upon which all nations' security and prosperity depend. The United States will continue to help safeguard access, promote security, and ensure the sustainable use of resources in these domains. These efforts require strong multilateral cooperation, enhanced domain awareness and monitoring, and the strengthening of international norms and standards.

We must work together to ensure the constant flow of commerce, facilitate safe and secure air travel, and prevent disruptions to critical communications. We must also safeguard the sea, air, and space domains from those who would deny access or use them for hostile purposes. This includes keeping strategic straits and vital sea lanes open, improving the early detection of emerging maritime threats, denying adversaries hostile use of the air domain, and ensuring the responsible use of space. As one key effort in the sea domain, for example, we will pursue ratification of the United Nations Convention on the Law of the Sea.

Many of these goals are equally applicable to cyberspace. While cyberspace relies on the digital infrastructure of individual countries, such infrastructure is globally connected, and securing it requires global cooperation. We will push for the recognition of norms of behavior in cyberspace, and otherwise work with global partners to ensure the protection of the free flow of information and our continued access. At all times, we will continue to defend our digital networks from intrusion and harmful disruption.

Arctic Interests: The United States is an Arctic Nation with broad and fundamental interests in the Arctic region, where we seek to meet our national security needs, protect the environment, responsibly manage resources, account for indigenous communities, support scientific research, and strengthen international cooperation on a wide range of issues.

IV. Conclusion

> "It's easy to forget that, when this war began, we were united, bound together by the fresh memory of a horrific attack and by the determination to defend our homeland and the values we hold dear. I refuse to accept the notion that we cannot summon that unity again. I believe with every fiber of my being that we, as Americans, can still come together behind a common purpose, for our values are not simply words written into parchment. They are a creed that calls us together and that has carried us through the darkest of storms as one nation, as one people."
>
> —President Barack Obama, West Point, New York, December 2, 2009

This strategy calls for a comprehensive range of national actions, and a broad conception of what constitutes our national security. Above all, it is about renewing our leadership by calling upon what is best about America—our innovation and capacity; our openness and moral imagination.

Success will require approaches that can be sustained and achieve results. One of the reasons that this nation succeeded in the second half of the 20th century was its capacity to pursue policies and build institutions that endured across multiple Administrations, while also preserving the flexibility to endure setbacks and to make necessary adjustments. In some instances, the United States has been able to carry forward this example in the years since the Cold War. But there are also many open questions, unfinished reforms, and deep divisions—at home and abroad—that constrain our ability to advance our interests and renew our leadership.

To effectively craft and implement a sustainable, results-oriented national security strategy, there must be effective cooperation between the branches of government. This Administration believes that we are strong when we act in line with our laws, as the Constitution itself demands. This Administration is also committed to active consultation with Congress, and welcomes robust and effective oversight of its national security policies. We welcome Congress as a full partner in forging durable solutions to tough challenges, looking beyond the headlines to take a long view of America's interests. And we encourage Congress to pursue oversight in line with the reforms that have been enacted through legislation, particularly in the years since 9/11.

The executive branch must do its part by developing integrated plans and approaches that leverage the capabilities across its departments and agencies to deal with the issues we confront. Collaboration across the government—and with our partners at the state, local, and tribal levels of government, in industry, and abroad—must guide our actions.

This kind of effective cooperation will depend upon broad and bipartisan cooperation. Throughout the Cold War, even as there were intense disagreements about certain courses of action, there remained a belief that America's political leaders shared common goals, even if they differed about how to reach them. In today's political environment, due to the actions of both parties that sense of common purpose is at times lacking in our national security dialogue. This division places the United States at a strategic disadvantage. It sets back our ability to deal with difficult challenges and injects a sense of anxiety and polarization into our politics that can affect our policies and our posture around the world. It must be replaced by a renewed sense of civility and a commitment to embrace our common purpose as Americans.

Americans are by nature a confident and optimistic people. We would not have achieved our position of leadership in the world without the extraordinary strength of our founding documents and the capability and courage of generations of Americans who gave life to those values—through their service, through their sacrifices, through their aspirations, and through their pursuit of a more perfect union. We see those same qualities today, particularly in our young men and women in uniform who have served tour after tour of duty to defend our nation in harm's way, and their civilian counterparts.

This responsibility cannot be theirs alone. And there is no question that we, as a nation, can meet our responsibility as Americans once more. Even in a world of enormous challenges, no threat is bigger than the American peoples' capacity to meet it, and no opportunity exceeds our reach. We continue to draw strength from those founding documents that established the creed that binds us together. We, too, can demonstrate the capability and courage to pursue a more perfect union and—in doing so—renew American leadership in the world.

In: U.S. National Security: A New Strategy for... ISBN: 978-1-61761-436-1
Editors: Belinda M. White

Chapter 2

THE NATIONAL SECURITY COUNCIL: AN ORGANIZATIONAL ASSESSMENT*

Richard A. Best, Jr.

SUMMARY

The National Security Council (NSC) was established by statute in 1947 to create an inter-departmental body to advise the President with respect to the integration of domestic, foreign, and military policies relating to the national security so as to enable the military services and the other departments and agencies of the Government to cooperate more effectively in matters involving the national security. Currently, statutory members of the Council are the President, Vice President, the Secretary of State, and the Secretary of Defense; but, at the President's request, other senior officials participate in NSC deliberations. The Chairman of the Joint Chiefs of Staff and the Director of National Intelligence are statutory advisers. In 2007 the Secretary of Energy was added to the NSC membership.

The President clearly holds final decision-making authority in the executive branch. Over the years, however, the NSC staff has emerged as a major factor in the formulation (and at times in the implementation) of

* This is an edited, reformatted and augmented edition of a United States Congressional Research Service publication, Report RL30840, dated June 8, 2009

national security policy. Similarly, the head of the NSC staff, the National Security Adviser, has played important, and occasionally highly public, roles in policymaking. This chapter traces the evolution of the NSC from its creation to the present.

The organization and influence of the NSC have varied significantly from one Administration to another, from a highly structured and formal system to loose-knit teams of experts. It is universally acknowledged that the NSC staff should be organized to meet the particular goals and work habits of an incumbent President. The history of the NSC provides ample evidence of the advantages and disadvantages of different types of policymaking structures.

Congress enacted the statute creating the NSC and has altered the character of its membership over the years. Congress annually appropriates funds for its activities, but does not, routinely, receive testimony on substantive matters from the National Security Adviser or from NSC staff. Proposals to require Senate confirmation of the Security Adviser have been discussed but not adopted.

The post-Cold War world has posed new challenges to NSC policymaking. Some argue that the NSC should be broadened to reflect an expanding role of economic, environmental, and demographic issues in national security policymaking. The Clinton Administration created a National Economic Council tasked with cooperating closely with the NSC on international economic matters. In the wake of the 9/11 attacks, the George W. Bush Administration established a Homeland Security Council. The Obama Administration has combined the staffs of the Homeland Security Council and the National Security Council, while retaining the two positions of National Security Adviser and Homeland Security Adviser. Although the latter has direct access to the President, the incumbent is to organizationally report to the National Security Adviser.

INTRODUCTION

The National Security Council (NSC) has been an integral part of U.S. national security policymaking since 1947. Of the various organizations in the Executive Office of the President that have been concerned with national security matters, the NSC is the most important and the only one established by statute. The NSC lies at the heart of the national security apparatus, being the highest coordinative and advisory body within the Government in this area aside from the President's Cabinet. The Cabinet has no statutory role, but the NSC does.

This study reviews the organizational history of the NSC and other related components of the Executive Office and their changing role in the national security policy process. It is intended to provide information on the NSC's development as well as subsequent usage. This study is not intended to be a comprehensive organizational history of all components of the national security policy process nor of the process itself as a whole. Moreover, the high sensitivity and security classification of the NSC's work and organization limit available sources. It is also important to keep in mind the distinction between the NSC's statutory membership (i.e., the President, Vice President, Secretary of State, Secretary of Defense, and Secretary of Energy) and its staff (i.e., the National Security Adviser and his assistants). These two groups have very different roles and levels of influence.

PRE-NSC COORDINATION METHODS

The Need for Interdepartmental Coordination

Successful national security policymaking is based on careful analysis of the international situation, including diplomatic, economic, intelligence, military, and morale factors. Based on a comprehensive assessment, effective government leaders attempt to attain their goals by selecting the most appropriate instrument of policy, whether it is military, diplomatic, economic, based on the intelligence services, or a combination of more than one. Although this approach has been an ideal throughout the history of international relations, prior to World War II, U.S. Presidents, focused primarily on domestic matters, and lacked organizational support to integrate national security policies. They relied instead on *ad hoc* arrangements and informal groups of advisers. However, in the early 1940s, the complexities of global war and the need to work together with allies led to more structured processes of national security decisionmaking to ensure that the efforts of the State, War, and Navy Departments were focused on the same objectives. There was an increasingly apparent need for an organizational entity to support the President in looking at the multiplicity of factors, military and diplomatic, that had to be faced during wartime and in the early postwar months when crucial decisions had to be made regarding the future of Germany and Japan and a large number of other countries.

Given continuing worldwide responsibilities in the postwar years that involved active diplomacy, sizable military forces, sophisticated intelligence agencies, in addition to economic assistance in various forms, the United States established organizational mechanisms to analyze the international environment, identify priorities, and recommend appropriate policy options. Four decades later, the end of the Cold War saw the emergence of new international concerns, including transnational threats such as international terrorism and drug trafficking, that have continued to require the coordination of various departments and agencies concerned with national security policies.

Past Modes of Policy Coordination

Coordinative mechanisms to implement policy are largely creations of the Executive Branch, but they directly influence choices that Congress may be called upon to support and fund. Congress thus takes interest in the processes by which policies and the roles of various participants are determined. Poor coordination of national security policy can result in calls for Congress to take actions that have major costs, both international and domestic, without the likelihood of a successful outcome. Effective coordination, on the other hand, can mean the achievement of policy goals with minimal losses of human lives while providing the opportunity to devote material resources to other needs.

Throughout most of the history of the United States, until the twentieth century, policy coordination centered on the President, who was virtually the sole means of such coordination. The Constitution designates the President as Commander-in-Chief of the armed forces (Article II, Section 2) and grants him broad powers in the areas of foreign affairs (Article II, Section 2), powers that have expanded considerably in the twentieth century through usage. Given limited U.S. foreign involvements for the first hundred or so years under the Constitution, the small size of the armed forces, the relative geographic isolation of the Nation, and the absence of any proximate threat, the President, or his executive agents in the Cabinet, provided a sufficient coordinative base.

However, the advent of World War I, which represented a modern, complex military effort involving broad domestic and international coordination, forced new demands on the system that the President alone could not meet. In 1916, the Council of National Defense was established by statute (Army Appropriation Act of 1916). It reflected proposals that went back to

1911 and consisted of the Secretaries of War, Navy, Interior, Agriculture, Commerce and Labor. The statute also allowed the President to appoint an advisory commission of outside specialists to aid the Council.[1] The Council of National Defense was intended as an economic mobilization coordinating group, as reflected by its membership—which excluded the Secretary of State. His inclusion would have given the Council a much wider coordinative scope. Furthermore, the authorizing statute itself limited the role of the Council basically to economic mobilization issues. The Council of National Defense was disbanded in 1921, but it set a precedent for coordinative efforts that would be needed in World War II.

The President remained the sole national security coordinator until 1938, when the prewar crisis began to build in intensity, presenting numerous and wide-ranging threats to the inadequately armed United States. The State Department, in reaction to reports of Axis activities in Latin America, proposed that interdepartmental conferences be held with War and Navy Department representatives. In April 1938, Secretary of State Cordell Hull, in a letter to President Franklin Roosevelt, formally proposed the creation of a standing committee made up of the second ranking officers of the three departments, for purposes of liaison and coordination. The President approved this idea, and the Standing Liaison Committee, or Liaison Committee as it was also called, was established, the members being the Under Secretary of State, the Chief of Staff of the Army, and the Chief of Naval Operations. The Standing Liaison Committee was the first significant effort toward interdepartmental liaison and coordination, although its work in the area was limited and uneven. The Liaison Committee largely concentrated its efforts on Latin American problems, and it met irregularly. Although it did foster some worthwhile studies during the crisis following the fall of France, it was soon superseded by other coordinative modes. It was more a forum for exchanging information than a new coordinative and directing body.[2]

An informal coordinating mechanism, which complemented the Standing Liaison Committee, evolved during the weekly meetings established by Secretary of War Henry L. Stimson, who took office in June 1940. Stimson arranged for weekly luncheons with his Navy counterpart, Frank Knox, and Cordell Hull, but these meetings also did not fully meet the growing coordinative needs of the wartime government.

In May 1940 President Roosevelt used the precedent of the 1916 statute and established the National Defense Advisory Council (NDAC), composed of private citizens with expertise in specific economic sectors.[3] As with the earlier Council of National Defense, NDAC was organized to handle problems of economic mobilization; and by the end of the year it had given way to another organization in a succession of such groups.

During the war, there were a number of interdepartmental committees formed to handle various issues, and, while these did help achieve coordination, they suffered from two problems. First, their very multiplicity was to some degree counter-productive to coordination, and they still represented a piecemeal approach to these issues. Second, and more important, these committees in many cases were not advising the President directly, but were advising his advisers. Although their multiplicity and possible overlapping fit Roosevelt's preferred working methods, they did not represent coordination at the top. Roosevelt ran the war largely through the Joint Chiefs of Staff (JCS), who were then an *ad hoc* and *de facto* group, and through key advisers such as Harry Hopkins and James F. Byrnes, and via his own personal link with British Prime Minister Winston Churchill.

The weekly meetings arranged by Stimson evolved, however, into a significant coordinative body by 1945, with the formal creation of the State, War, Navy Coordinating Committee (SWNCC). SWNCC had its own secretariat and a number of regional and topical subcommittees; its members were assistant secretaries in each pertinent department. The role of SWNCC members was to aid their superiors "on politico-military matters and [in] coordinating the views of the three departments on matters in which all have a common interest, particularly those involving foreign policy and relations with foreign nations...." SWNCC was a significant improvement in civilian-military liaison, and meshed well with the JCS system; it did not, however, concern itself with fundamental questions of national policy during the early months of the Cold War.[4] SWNCC operated through the end of the war and beyond, becoming SANACC (State, Army, Navy, Air Force Coordinating Committee) after the National Security Act of 1947. It was dissolved in 1949, by which time it had been superseded by the NSC.

The creation of SWNCC, virtually at the end of the war, and its continued existence after the surrender of Germany and Japan reflected the growing awareness within the Federal Government that better means of coordination were necessary. The World War II system had largely reflected the preferred working methods of President Roosevelt, who relied on informal consultations with various advisers in addition to the JCS structure. However, the complex demands of global war and the post-war world rendered this system inadequate, and it was generally recognized that a return to the simple and limited prewar system would not be possible if the United States was to take on the responsibilities thrust upon it by the war and its aftermath.

THE CREATION OF THE NSC[5]

Introduction

The NSC was not created independently, but rather as one part of a complete restructuring of the entire national security apparatus, civilian and military, including intelligence efforts, as accomplished in the National Security Act of 1947. Thus, it is difficult to isolate the creation of the NSC from the larger reorganization, especially as the NSC was much less controversial than the unification of the military and so attracted less attention.

Proposals

As early as 1943, General George C. Marshall, Army Chief of Staff, had proposed that the prospect of a unified military establishment be assessed. Congress first began to consider this idea in 1944, with the Army showing interest while the Navy was opposed. At the request of the Navy these investigations were put off until 1945, although by then it was clear to Secretary of the Navy James Forrestal that President Truman, who had come to the White House upon the death of President Roosevelt in April 1945, favored some sort of reorganization. Forrestal believed that outright opposition would not be a satisfactory Navy stance. He also realized that the State Department had to be included in any new national security apparatus. Therefore, he had Ferdinand Eberstadt, a leading New York attorney and banker who had served in several high-level Executive Branch positions, investigate the problem.[6]

With respect to the formation of the NSC, the most significant of the three questions posed by Forrestal to Eberstadt, was:

> What form of postwar organization should be established and maintained to enable the military services and other governmental departments and agencies most effectively to provide for and protect our national security?

Eberstadt's response to this question covered the military establishment, where he favored three separate departments and the continuation of the JCS, as well as the civilian sphere, where he suggested the formation of two new major bodies "to coordinate all these [civilian and military] elements." These two bodies he called the National Security Council (NSC), composed of the President, the Secretaries of State and of the three military departments, the JCS "in attendance," and the chairman of the other new body, the National Security Resources Board (NSRB). Eberstadt also favored the creation of a Central Intelligence Agency (CIA) under the NSC.[7]

Eberstadt's recommendations clearly presaged the eventual national security apparatus, with the exception of a unified Department of Defense. Furthermore, it was a central point in Forrestal's plans for holding the proposed reorganization to Navy desires, bringing in the State Department, as he desired, and hopefully obviating the need for some coalescence of the military services. The NSC was also a useful negotiating point for Forrestal with the Army, as Eberstadt had described one of its functions as being the "building up [of] public support for clear-cut, consistent, and effective foreign and military policies." This would appeal to all the service factions as they thought back on the lean and insecure prewar years.[8]

War-Navy negotiations over the shape of the reorganization continued throughout 1946 and into 1947. However, some form of central coordination, for a while called the Council of Common Defense, was not one of the contentious issues. By the end of May 1946, agreement had been reached on this and several other points, and by the end of the year the two sides had agreed on the composition of the new coordinative body.[9]

Congressional Consideration

The creation of the NSC was one of the least controversial sections of the National Security Act and so drew little attention in comparison with the basic concept of a single military department, around which most of the congressional debate centered.

The concept of a regular and permanent organization for the coordination of national security policy was as widely accepted in Congress as in the Executive. When the NSC was considered in debate, the major issues were the mechanics of the new organization, its membership, assurances that it would be a civilian organization and would not be dominated by the new Secretary of the National Military Establishment, and whether future positions on the NSC would be subject to approval by the Senate.[10]

The NSC as Created in 1947

The NSC was created by the National Security Act, which was signed by the President on July 26, 1947. The NSC appears in Section 101 of Title I, Coordination for National Security, and its purpose is stated as follows:

> (a) ... The function of the Council shall be to advise the President with respect to the integration of domestic, foreign, and military policies relating to the national security so as to enable the military services and the other departments and agencies of the Government to cooperate more effectively in matters involving the national security.
>
> (b) In addition to performing such other functions as the President may direct, for the purpose of more effectively coordinating the policies and functions of the departments and agencies of the Government relating to the national security, it shall, subject to the direction of the President, be the duty of the Council
>
> (1) to assess and appraise the objectives, commitments, and risks of the United States in relation to our actual and potential military power, in the interest of national security, for the purpose of making recommendations to the President in connection there with; and

(2) to consider policies on matters of common interest to the departments and agencies of the Government concerned with the national security, and to make recommendations to the President in connection therewith. . . .

> (d) The Council shall, from time to time, make such recommendations, and such other reports to the President as it deems appropriate or as the President may require.[11]

The following officers were designated as members of the NSC: the President; the Secretaries of State, Defense, Army, Navy, and Air Force; and the Chairman of the National Security Resources Board. The President could also designate the following officers as members "from time to time:" secretaries of other executive departments and the Chairmen of the Munitions Board and the Research and Development Board. Any further expansion required Senate approval. The NSC was provided with a staff headed by a civilian executive secretary, appointed by the President.

The National Security Act also established the Central Intelligence Agency under the NSC, but the Director of Central Intelligence (DCI) was not designated as an NSC member. The act also created a National Military

Establishment, with three executive departments (Army, Navy, and Air Force) under a Secretary of Defense.

Implicit in the provisions of the National Security Act was an assumption that the NSC would have a role in ensuring that the U.S. industrial base would be capable of supporting national security strategies. The Chairman of the National Security Resources Board, set up by the same act to deal directly with industrial base and civilian mobilization issues, was provided a seat on the NSC. Over the years, however, these arrangements proved unsatisfactory and questions of defense mobilization and civil defense were transferred to other federal agencies and the membership of the NSC was limited to the President, Vice President, the Secretary of State and the Secretary of Defense.[12] Thus, the need for a coordinative entity that had initially been perceived to center on economic mobilization issues during World War I had evolved to one that engaged the more permanent themes of what had come to be known as national security policy.

The creation of the NSC was a definite improvement over past coordinative methods and organization, bringing together as it did the top diplomatic, military, and resource personnel with the President. The addition of the CIA, subordinate to the NSC, also provided the necessary intelligence and analyses for the Council so that it could keep pace with events and trends. The changeable nature of its organization and its designation as an advisory body to the President also meant that the NSC was a malleable organization, to be used as each President saw fit. Thus, its use, internal substructure, and ultimate effect would be directly dependent on the style and wishes of the President.

The National Security Council, 1947-2009

The Truman NSC, 1947-1953

Early Use. The NSC first met on September 26, 1947. President Truman attended the first session, but did not attend regularly thereafter, thus emphasizing the NSC's advisory role. In his place, the President designated the Secretary of State as chairman, which also was in accord with the President's view of the major role that the State Department should play. Truman viewed the NSC as a forum for studying and appraising problems and making recommendations, but not one for setting policy or serving as a centralized office to coordinate implementation.

The NSC met irregularly for the first 10 months. In May 1948, meetings twice a month were scheduled, although some were canceled, and special sessions were convened as needed.

The Hoover Commission. The first review of NSC operations came in January 1949 with the report of the Hoover Commission (the Commission on Organization of the Executive Branch of the Government), which found that the NSC was not fully meeting coordination needs, especially in the area of comprehensive statements of current and long-range policies.[13]

The Hoover Commission recommended that better working-level liaison between the NSC and JCS be developed, that the Secretary of Defense become an NSC member, replacing the service secretaries, and that various other steps be taken to clarify and tighten roles and liaison.

1949 Amendments. In January 1949, President Truman directed the Secretary of the Treasury to attend all NSC meetings. In August 1949, amendments to the National Security Act were passed (P.L. 81-216), changing the membership of the NSC to consist of the following officers: the President, Vice President, Secretaries of State and Defense, and Chairman of the National Security Resources Board. This act also designated the JCS as "the principal military advisers to the President," thus opening the way for their attendance, beginning in 1950, even though the Service Secretaries were excluded. In August 1949, by Reorganization Plan No. 4, the NSC also became part of the Executive Office of the President, formalizing a *de facto* situation.

Subsequent Usage and Evaluation. The outbreak of the Korean War in June 1950 brought greater reliance on the NSC system. The President ordered weekly meetings and specified that all major national security recommendations be coordinated through the NSC and its staff. Truman began presiding regularly, chairing 62 of the 71 meetings between June 1950 and January 1953. The NSC became to a much larger extent the focus of national security decisionmaking. Still, the NSC's role remained limited. Truman continued to use alternate sources of information and advice.[14] As one scholar has concluded:

Throughout his administration Truman's use of the NSC process remained entirely consistent with his views of its purpose and value. The president and his secretary of state remained completely responsible for foreign policy. Once policy decisions were made, the NSC was there to advise the president on matters requiring specific diplomatic, military, and intelligence coordination.[15]

The Eisenhower NSC, 1953-1961

President Dwight Eisenhower, whose experience with a well-ordered staff was extensive, gave new life to the NSC. Under his Administration, the NSC staff was institutionalized and expanded, with clear lines of responsibility and authority, and it came to closely resemble Eberstadt's original conception as the President's principal arm for formulating and coordinating military, international, and internal security affairs. Meetings were held weekly and, in addition to Eisenhower himself and the other statutory members, participants often included the Secretary of the Treasury, the Budget Director, the Chairman of the JCS, and the Director of Central Intelligence.

Organizational Changes. In his role as chairman of the NSC, Eisenhower created the position of Special Assistant for National Security Affairs,[16] who became the supervisory officer of the NSC, including the Executive Secretary. The Special Assistant—initially Robert Cutler, a banker who had served under Stimson during World War II—was intended to be the President's agent on the NSC, not an independent policymaker in his own right, and to be a source of advice.[17]

Eisenhower established two important subordinate bodies: the NSC Planning Board, which prepared studies, policy recommendations, and basic drafts for NSC coordination, and the Operations Coordinating Board, which was the coordinating and integrating arm of the NSC for all aspects of the implementation of national security policy.

By the end of the Eisenhower Administration, the NSC membership had changed slightly. The National Security Resources Board had been abolished by Reorganization Plan No. 3 in June 1953, and this vacancy was then filled by the Director of the Office of Civil and Defense Mobilization.

In 1956, President Eisenhower, partly in response to recommendations of the second Hoover Commission on the Organization of the Executive Branch of Government, also established the Board of Consultants on Foreign

Intelligence Activities in the Executive Office. This board was established by Executive Order 10656 and was tasked to provide the President with independent evaluations of the U.S. foreign intelligence effort. The Board of Consultants lapsed at the end of the Eisenhower Administration, but a similar body, the President's Foreign Intelligence Advisory Board (PFIAB), was created by President Kennedy after the Bay of Pigs failure. PFIAB was itself abolished in 1977, but was resurrected during the Reagan Administration in 1981. Members are selected by the President and serve at his discretion.

Evaluation. The formal structure of the NSC under Eisenhower allowed it to handle an increasing volume of matters. Its work included comprehensive assessments of the country's basic national security strategy, which were designed to serve as the basis for military planning and foreign policymaking. The complexity of NSC procedures under Eisenhower and its lengthy papers led to charges that quantity was achieved at the expense of quality and that the NSC was too large and inflexible in its operations. Critics alleged that it was unable to focus sufficiently on major issue areas.[18] Some observers also held that NSC recommendations were often compromises based on the broadest mutually acceptable grounds from all the agencies involved, leading to a noticeable lack of innovative national security ideas. The Eisenhower NSC did, nonetheless, establish national security policies that were accepted and implemented throughout the Government and that laid the basis for sustained competition with the Soviet Union for several decades.

It may be that the NSC process became overly bureaucratic towards the end of the Eisenhower Administration, perhaps affected by the President's declining health. Hearings by the Senate Government Operations Committee in 1960-61, led by Senator Henry Jackson, produced proposals for a substantial reorientation of this "over-institutionalized" structure, and its replacement by a smaller, less formal NSC that would offer the President a clear choice of alternatives on a limited number of major problems.[19]

Some scholars have noted that Eisenhower himself found the lengthy NSC procedures burdensome and argue that many key decisions were made in the Oval Office in the presence of only a few advisers. Nonetheless, Eisenhower saw the NSC process as one which produced a consensus within the Administration which would lead to effective policy implementation. According to this view, the process was largely one of education and clarification.[20] A recent analysis has concluded, that NSC meetings

> brought Eisenhower's thinking into sharper focus by forcing him to weigh it against a range of alternatives that were presented and defended by individuals whose opinions the president took seriously and whose exposure to requisite information and expertise he assured. These individuals, in turn, were educated about the problems in the same way as Eisenhower.[21]

The Kennedy NSC, 1961-1963

President John Kennedy, who did not share Eisenhower's preference for formal staff procedures, accepted many of the recommendations of the Jackson Committee and proceeded to dismantle much of the NSC structure, reducing it to its statutory base. Staff work was carried out mainly by the various departments and agencies, and personal contacts and *ad hoc* task forces became the main vehicles for policy discussion and formulation. The NSC was now one among many sources of advice.

Kennedy's National Security Adviser, McGeorge Bundy, played an important policy role directly under the President. The nature of this position was no longer that of a "neutral keeper of the machinery"; for the first time, the Adviser emerged in an active policymaking role, in part because of the absence of any definite NSC process that might preoccupy him.[22]

Kennedy met regularly with the statutory NSC members and the DCI, but not in formal NSC sessions. Studies and coordination were assigned to specific Cabinet officers or subordinates in a system that placed great emphasis on individual responsibility, initiative and action. The Secretary of State, Dean Rusk, was initially seen as the second most important national security official in the President's plans, and Kennedy indicated that he did not want any other organizations interposed between him and Rusk. However, Kennedy came to be disappointed by the State Department's inability or unwillingness to fill this role as the leading agency in national security policy.[23]

At the beginning of the Kennedy Administration, the NSC was reportedly cut from seventy-one to forty-eight and "in place of weighty policy papers, produced at regular intervals, Bundy's staff would produce crisp and timely National Security Action Memoranda (NSAMs). The new name signified the premium that would be placed on 'action' over 'planning.'"[24] With an emphasis on current operations and crisis management, special *ad hoc* bodies came into use. The outstanding example of this was the Executive Committee (ExCom), formed in October 1962 during the Cuban Missile Crisis, which orchestrated the U.S. response to Soviet moves to introduce missiles in Cuba.

Organizational Changes. Kennedy added the Director of the Office of Emergency Planning to the NSC, replacing the Director of the Office of Civil and Defense Mobilization. It was planned that the new appointee would fill the role originally envisioned for the National Security Resources Board in coordinating emergency management of resources.

The Planning Board and the Operations Control Board were both abolished (by Executive Order 10920) in order to avoid the Eisenhower Administration's distinction between planning and operations. The NSC staff was reduced, and outside policy experts were brought in. Bundy noted that they were all staff officers:

> Their job is to help the President, not to supersede or supplement any of the high officials who hold line responsibilities in the executive departments and agencies. Their task is that of all staff officers: to extend the range and enlarge the direct effectiveness of the man they serve. Heavy responsibilities for operation, for coordination, and for diplomatic relations can be and are delegated to the Department of State. Full use of all the powers of leadership can be and is expected in other departments and agencies. There remains a crushing burden of responsibility, and of sheer work, on the President himself; there remains also the steady flow of questions, of ideas, of executive energy which a strong President will give off like sparks. If his Cabinet officers are to be free to do their own work, the President's work must be done—to the extent that he cannot do its himself—by staff officers under his direct oversight. But this is, I repeat, something entirely different from the interposition of such a staff between the President and his Cabinet officers.[25]

Evaluation. Some critics attacked the informality of the system under Kennedy, arguing that it lacked form and direction, as well as coordination and control, and that it emphasized current developments at the expense of planning. As noted, Kennedy himself was disappointed by the State Department, on which he had hoped to rely. In retrospect, Kennedy's system was designed to serve his approach to the presidency and depended upon the President's active interest and continuous involvement. Some critics, both at the time and subsequently, have suggested that the informal methods that the Kennedy Administration adopted contributed to the Bay of Pigs debacle and the confusion that surrounded U.S. policy in the coup against President Diem of South Vietnam in 1963.

The Johnson NSC, 1963-1969

President Lyndon Johnson's sudden accession to power, the need for a show of continuity, and pressures from the upcoming Presidential election all forced Johnson, at least until 1965, to rely heavily on Kennedy's system and personnel, especially as Johnson was less familiar with national security than domestic affairs.

Organizational Changes. Johnson, like Truman, sought out advice from a number of sources other than the NSC and its member departments, although he relied heavily on the Secretaries of State and Defense, Dean Rusk and Robert McNamara.

The institutional system that evolved under Johnson depended heavily on the ability of the State Department to handle the planning and coordination process. This system came about from a study headed by General Maxwell Taylor in 1966 that led to National Security Action Memorandum (NSAM) 341 which concluded that it was necessary to enhance the State Department's role in the policy process and to improve "country team expertise" in Washington, which was felt to be far below that in the various embassies. NSAM 341 led to a new system of interagency committees. The most important of these was the Senior Interdepartmental Group (SIG), whose members were: the Under Secretary of State, Deputy Secretary of Defense, Administrator of the Agency for International Development, DCI, JCS Chairman, Director of the U.S. Information Agency, and the National Security Adviser. In support of the SIG were a number of Interdepartmental Regional Groups (IRGs), each headed by the appropriate Assistant Secretary of State.

Within the NSC itself, structure and membership remained what they had been under Kennedy (with the Office of Emergency Planning changing title to the Office of Emergency Preparedness in 1968), although the title Special Assistant to the President for National Security Affairs was shortened to Special Assistant when Walt W. Rostow replaced Bundy in 1966. This reflected the frequent diversion of the occupant of this position away from NSC affairs to more general concerns.

Evaluation. Johnson's NSC system barely existed as such. The role of the NSC staff was more restricted, and budget and personnel both declined. Key decisions, those especially regarding the war in Vietnam, were made during Tuesday lunches attended by the President, the Secretaries of State and Defense, and a few other invited officials.

Johnson's informal system was not a wholly successful replacement for the highly structured system developed in the Eisenhower Administration. The SIG/IRG system fulfilled neither old functions nor the objectives set forth in NSAM 341. Although this new structure was dominated by the State Department, there was little enthusiasm for the system as a whole on the part of the department's leadership. The State Department did not provide decisive leadership and settled for a system of consensus opinions. Vagueness as to authority in the SIG/IRG system reduced its effect on the bureaucracies. Moreover, there was an insufficient allocation of resources for staff support for the new organization. By 1969, the NSC existed largely in name. Johnson conferred constantly with a wide number of advisers within and outside government; while he respected institutional responsibilities, his own decisionmaking was an intensely personal process.

The Nixon NSC, 1969-1974

Experience in the Eisenhower Administration clearly had a formative effect on President Richard Nixon's approach to national security organization. Wanting to switch White House priorities from current operations and crisis management to long-range planning, Nixon revived the NSC. Nixon's NSC staff structure resembled Eisenhower's, with an emphasis on examining policy choices and alternatives, aiming for a number of clear options reaching the highest level, where they would be treated systematically and then effectively implemented. Nixon made it clear that he wanted distinct options presented to him from which he could choose, rather than consensus opinions requiring only acceptance or rejection. Nixon used an NSC framework similar to that set in place by Eisenhower but intended, as much as Kennedy, to give the NSC staff a powerful policy role.

Organizational Changes. While adopting the basic form of the Eisenhower NSC, Nixon streamlined its procedures.[26] The position of Assistant for National Security Affairs was revived, and Henry Kissinger, a Harvard professor and occasional government adviser, was named to fill it. NSC meetings were limited to the statutory members, with Kissinger and the JCS Chairman also sitting in and the DCI attending for intelligence matters. In January 1973, the Office of Emergency Preparedness was abolished along with the NSC seat that originally had belonged to the Chairman of the National Security Resources Board.

Six interdepartmental groups, similar to Johnson's IRGs, formed the NSC's support network, preparing basic studies and developing policy options. However, the influence of the State Department was reduced, and Kissinger's influence soon predominated. Four major new bodies were created:

- Washington Special Action Group (WASAG): headed by Kissinger and designed to handle contingency planning and crises.
- NSC Intelligence Committee: chaired by Kissinger and responsible for providing guidance for national intelligence needs and continuing evaluations of intelligence products.
- Defense Program Review Committee: chaired by Kissinger and designed to achieve greater integration of defense and domestic considerations in the allocation of natural resources. This committee was intended to allow the President, through the NSC, to gain greater control over the defense budget and its implications and policy requirements. As a result of opposition by Defense Secretary Melvin Laird, its role was, however, significantly circumscribed.
- Senior Policy Review Group: chaired by Kissinger, this group directed and reviewed policy studies and also served as a top level deliberative body.

This system had two principal objectives: the retention of control at the top, and the development of clear alternative choices for decisionmakers.

Evaluation. Most of the criticism of the Nixon NSC centered on the role played by Kissinger. His position in a number of the key committees gave him control over virtually the entire NSC apparatus, leading to charges that the system, for all its efficiency, now suffered from over-centralization, and later from domination by one man.

During Nixon's first term, Kissinger competed with the State Department for control of foreign policy, and soon overshadowed Secretary of State William Rogers. Critics felt that Kissinger stifled dissent within the NSC and the rest of the national security apparatus. Kissinger's venture into "shuttle diplomacy" and the unique circumstances of the Watergate scandal further emphasized his key role. Kissinger's accession to Rogers' position in September 1973, while retaining his National Security Council post, brought renewed criticism of his role. The direct involvement of the NSC Adviser in diplomatic negotiations set a precedent that some observers have criticized as undercutting the established responsibilities of the State Department and as an attempt to orchestrate national security policy beyond the reach of congressional oversight.

Kissinger's predominance derived from his unique intellectual abilities, skill at bureaucratic maneuvering, and the support of a President determined to act boldly in international affairs without being restrained by bureaucratic or congressional inhibitions. It was achieved at a time of profound political differences over foreign policy in which Administration and congressional goals were, on occasion, diametrically opposed. However, under President Nixon, the NSC was restored to a central role in the policy process, acting as the major vehicle and conduit for the formation of national security policy.

The Ford NSC, 1974-1977

President Gerald Ford, who inherited his predecessor's NSC, took no major steps to change the system per se, although Kissinger was replaced as National Security Adviser by Air Force Lt. General Brent Scowcroft in November 1975. The national security policy process continued to be dominated by Kissinger, who retained his position as Secretary of State, an indication of the preeminence he had achieved, as well as a reflection of Ford's limited experience in the conduct of foreign policy prior to his sudden accession.

In June 1975, the Commission on the Organization of the Government for the Conduct of Foreign Policy, also known as the Murphy Commission, issued a report on ways to more effectively formulate and implement foreign policy. Its recommendations dealt in part with the Executive Office of the President and the NSC structure.[27]

Implicitly criticizing the expansive role of the NSC staff under Kissinger, the Commission recommended that only the President should have line responsibility in the White House; that staff officials should not themselves issue directives to departmental officials; that, in the future, the National Security Adviser have no other official responsibilities; that the Secretary of the Treasury be made a statutory member of the NSC and that the NSC's scope be expanded to include major international economic policy issues; and that senior officials concerned with domestic policy be invited to NSC meetings when issues with domestic implications were discussed.

The Commission also considered general alternative structures and pointed out their basic advantages and disadvantages. It also noted that,

> Policymaking is not a branch of mechanics; however wisely designed or carefully utilized, no machinery is adequate to assure its results. The selective use of various mechanisms and forums in ways which fit the particular issues, positions, and personalities involved is as much a part of the President's responsibility as is the necessity, finally, to decide the substantive issues.[28]

There were no immediate steps taken to implement the Report's recommendations.

Organizational Changes. In February 1976, President Ford issued Executive Order 11905 reorganizing the intelligence community, in response to ongoing investigations in that area. This order, among other things, reaffirmed the NSC's overall policy control over the foreign intelligence community. Some changes were made in the NSC sub-structure, including the abolition of the NSC Intelligence Committee. The so-called 40 Committee of the NSC, which was responsible for covert operations and certain sensitive foreign intelligence operations, was replaced by the Operations Advisory Group. This Executive Order also created the Intelligence Oversight Board in the Executive Office (subsequently disbanded in 1993). It was composed of three civilians and was tasked with reviewing the propriety and legality of the intelligence agencies' operations. In December 1975, Ford vetoed a bill that would have made the Secretary of the Treasury a statutory member of the NSC, saying that the Treasury Secretary is invited to participate in NSC affairs having significant economic and monetary implications, but that there is no need to involve him in all NSC activities.[29]

Evaluation. These changes did not detract from the central role that the NSC had achieved under President Nixon. Kissinger's loss of his dual position did not seem to lessen his influence over the policy process, leading critics to charge that this change was largely cosmetic. The new National Security Adviser, Brent Scowcroft, had previously served as Kissinger's deputy on the NSC staff, and was unlikely to challenge Kissinger's pre-eminence. The Ford NSC reflected the close relationship between the President and the Secretary of State, a relationship that itself became a source of controversy both in the Republican primaries of 1976 as well as the ensuing general election. Critics continued to maintain that the Ford Administration decisionmaking was secretive, impervious to congressional input, and out of touch with public opinion.

The Carter NSC, 1977-1981[30]

Under President Jimmy Carter, steps were taken to end the dominant role of the NSC staff and make it a more coequal and cooperating partner with the Departments of State and Defense. The NSC underwent a major reorganization in the new Administration.

Organizational Changes. Upon taking office in January 1977, President Carter issued a directive (PD-2) reorganizing the NSC staff. The avowed purpose of the reorganization was "to place more responsibility in the departments and agencies while insuring that the NSC, with my Assistant for National Security Affairs, continues to integrate and facilitate foreign and defense policy decisions."[31]

The number of NSC staff committees was reduced from seven to two, the Policy Review Committee (PRC) and the Special Coordination Committee (SCC). The functions of these two committees were as follows:

- Policy Review Committee: the PRC had responsibility for subjects which "fall primarily within a given department but where the subject also has important implications for other departments and agencies." Examples were "foreign policy issues that contain significant military or other interagency aspects; defense policy issues having international implications and the coordination of the annual Defense budget with foreign policy objectives; the preparation of a consolidated national intelligence budget and resource allocation for the Intelligence Community...; and those international economic issues pertinent to the U.S. foreign policy and security...."[32] Executive Order 12036 of January 24, 1978, added responsibility for the establishment of national foreign intelligence requirements and priorities, and periodic reviews and evaluations of national foreign intelligence products.[33] The Vice President, the Secretaries of State and Defense, and the Assistant for National Security Affairs were members of the PRC; the DCI and the Chairman of the JCS also attended. The Secretary of the Treasury, the Chairman of the Council of Economic Advisers, and other officials attended when pertinent topics were being considered. Appropriate Cabinet officers chaired the PRC in accordance with matters being considered; the DCI was chairman when the PRC considered intelligence matters as specified in E.O. 12036. NSC Interdepartmental Groups, which dealt with specific issues at the direction of the President, were under the PRC.

- Special Coordination Committee: the SCC dealt with "specific, cross-cutting issues requiring coordination in the development of options and the implementation of Presidential decisions." These included "oversight of sensitive intelligence activities ... arms control evaluation; ... and crisis management."[34] E.O. 12036 gave the SCC responsibility for sensitive foreign intelligence collection operations and counterintelligence.[35] The SCC thus replaced WASAG and the Operations Advisory Group. Unlike the PRC, the SCC was chaired by the Assistant for National Security Affairs; other members were the Vice President, the Secretaries of State and Defense, and the DCI—or their deputies—and other officials attended when appropriate. When intelligence "special activities" were being considered, the members had to attend, as had the Attorney General, the Chairman of the JCS, and the Director of the Office of Management and Budget (OMB); for counterintelligence activities, the Director of the FBI attended.[36]
- The initial emphasis of the NSC's role as a policy coordinator and "think tank" represented a clear reversal of the trend that had developed under Presidents Nixon and Ford. The staff of the NSC was reduced under the Carter Administration, and National Security Adviser Zbigniew Brzezinski established a number of regional and topical offices on the NSC staff that aimed at a more "collegial" approach to staff procedures.
- Although the PRC had a wider charter than the SCC, as a result of the growing importance of crisis management functions and the increasing influence of the National Security Adviser,[37] initiative passed to the SCC and there were fewer PRC meetings.
- Evaluation. A rumored rivalry between Brzezinski and Secretary of State Cyrus Vance was not publicly evident during the first year of the Carter Administration, but reports of differences between the two men later increased dramatically as senior Administration officials advised different responses to such questions as Soviet and Cuban activities in Africa and the Iranian hostage question. Towards the end of the Administration, differences between Vance and Brzezinski became pronounced and were widely perceived as contributing to weak and vacillating policies. Carter's Director of Central Intelligence Stansfield Turner later wrote:

> National Security Advisers and Secretaries of State and Defense had clashed before, notably under President Nixon when Henry Kissinger was the Adviser. But because Nixon tended to follow Kissinger's advice more often than not, there was no stalemate, and foreign policy moved ahead in innovative ways. However, Jimmy Carter vacillated between Brzezinski and Vance, and they often canceled each other out.[38]

Vance, who had strongly opposed the ill-fated effort to rescue the U.S. hostages in Iran, finally resigned and was succeeded by Senator Edmund Muskie in April 1980. Brzezinski's outspokenness and his public role in policymaking became an issue, and led to calls for Senate confirmation of NSC advisers and closer congressional oversight of the NSC staff.[39] There were also reports of infighting between Carter loyalists on the NSC staff and those who had worked for Vice President Walter Mondale, who had been given a major policy role.[40]

The Reagan NSC, 1981-1989

Campaigning for the presidency in 1980, Ronald Reagan criticized the divisions of the Carter Administration and promised to restore Cabinet leadership (as, in the 1976 campaign, he had criticized Henry Kissinger's predominant influence in the Ford Administration). Substituting Cabinet leadership for an active NSC proved, however, to be a significant challenge.

Organizational Changes. After extensive delays and bureaucratic infighting, President Reagan signed a Presidential directive (NSDD-2),[41] which enhanced the role of the State Department in national security policymaking and downgraded that of the National Security Adviser. The various NSC sub-committees were to be chaired by State, Defense, and CIA officials, not NSC staff. The Reagan NSC included three Senior Interagency Groups (SIGs)—one for foreign policy, chaired by the Deputy Secretary of State; one for defense, chaired by the Deputy Secretary of Defense; and one for intelligence, chaired by the Director of Central Intelligence. There were also regional and functional interagency groups, chaired by representatives of various Cabinet departments. Crisis management formally became the direct responsibility of the Vice President.[42]

This structure, however, had major limitations. Observers and participants portray an absence of orderly decisionmaking and uncertain lines of responsibility. As the Special Review Board (known as the Tower Board) appointed by the President to assess the proper role of the NSC system in the wake of the Iran-Contra revelations, pointedly noted:

> A President must at the outset provide guidelines to the members of the National Security Council, his National Security Adviser, and the National Security Council staff. These guidelines, to be effective, must include how they will relate to one another, what procedures will be followed, what the President expects of them. If his advisors are not performing as he likes, only the President can intervene.[43]

The Reagan Administration had a total of six National Security Advisers. Their history is poignant. The first, Richard Allen, did not have direct access to the President, but reported to him through Presidential Counselor Edwin Meese. Allen's tenure was brief; after accusations of influence peddling, he was replaced in January 1982 by Judge William Clark, a longtime Reagan associate who had served since the beginning of the Administration as Deputy Secretary of State. Clark, in turn, resigned in October 1983 to become Secretary of the Interior and his deputy, Robert McFarlane, became National Security Adviser. McFarlane was replaced in January 1986 by his deputy, Vice Admiral John Poindexter, and subsequently pleaded guilty to withholding information from Congress. Poindexter himself was relieved in the context of the Iran-Contra scandal in November 1986, and eventually went on trial for obstructing justice. An effort was made to restore NSC effectiveness under former Ambassador Frank Carlucci, who succeeded Poindexter in December 1986. When Carlucci was appointed Secretary of Defense, he was replaced by Army General Colin Powell in November 1987.

Evaluation. Until the arrival of Carlucci, the Reagan NSC structure lacked a strong, politically attuned National Security Adviser that had characterized Administrations since 1961. It also lacked the administrative structure that existed under Eisenhower, Nixon, Ford, and Carter. The absence of either influential NSC Advisers or effective administrative machinery has been seen by many critics as a major factor contributing to the Iran-Contra misadventures. Allowing NSC committees to be chaired by Cabinet officials tended to reduce the possibility that all sides of a given issue would be laid before the full NSC or the President. The Tower Board noted:

> Most presidents have set up interagency committees at both a staff and policy level to surface issues, develop options, and clarify choices. There has typically been a struggle for the chairmanship of these groups between the National Security Adviser and the NSC staff on the one hand, and the cabinet secretaries and department officials on the other.
>
> Our review of the operation of the present system and that of other administrations where committee chairmen came from the Departments has led us to the conclusion that the system generally operates better when the committees are chaired by the individual with the greatest stake in making the NSC system work.
>
> We recommend that the National Security Adviser chair the senior-level committees of the NSC system.[44]

The Reagan Administration, in its efforts to avoid the dominant influence wielded by previous NSC Advisers, fell victim to perpetual bureaucratic intrigues. The efforts of politically weak NSC Advisers, especially McFarlane and Poindexter, to undertake White House initiatives covertly over the strong opposition of senior Cabinet officials and congressional leaders called into question the basic competence of the Administration.

Another aspect of the Reagan NSC that came under heavy criticism was the involvement of NSC staff in covert actions. Although NSC staff efforts to manage certain crises, such as the capture of the *Achille Lauro* hijackers, were successful, the participation of NSC personnel, especially Lt. Col. Oliver North, in operations run apart from the traditional intelligence apparatus, including efforts to gain the release of American hostages and to supply Nicaraguan insurgents, has been widely censured. Such efforts have been criticized as undercutting the agencies with responsibilities for such operations and which are accountable to congressional oversight committees; secondly, failing to take full advantage of the professional expertise available to the Intelligence Community, and potentially involving the country in misguided ventures. The Iran-Contra Committee recommended that "the members and staff of the NSC not engage in covert actions."[45]

Reagan's final two NSC Advisers, Carlucci and Powell, brought a period of greater stability to NSC operations and both eschewed participation in covert actions. After Poindexter's departure, Carlucci created a Senior Review Group that he himself chaired and that was composed of statutory NSC members (besides the President and Vice President). He also established a Policy Review Group that was chaired by his deputy and composed of second-ranking officials of NSC agencies.

President Reagan's own role in the details of national security policymaking remains unclear. His policies on U.S.-Soviet relations, support for an aggressive struggle against international communism, and the need for strong military forces, including strategic defenses were well-known; such positions provided the overall goals for Administration officials. It is generally acknowledged, however, that unlike some of his predecessors, President Reagan did not himself engage in detailed monitoring of policy implementation. Some maintain that his NSC structure and the absence of strong NSC Advisers led directly to bureaucratic gridlock and ill-advised involvement of the NSC staff in covert actions. Others have concluded that the experience of the Reagan Administration demonstrates that a strong and efficient National Security Adviser and staff has become essential to national security policymaking, especially if the President himself does not provide detailed direction. The absence of such an Adviser, it is argued, will undermine the development and implementation of effective national security policies. Some subsequent historians, however, give Reagan higher marks for overall national security policy even if his NSC staff was often in flux.

The George H.W. Bush NSC, 1989-1993

The Bush Administration saw the return of Brent Scowcroft as National Security Adviser. His tenure was marked by the absence of public confrontations with Cabinet officers and a close working relationship with the President. National Security Directive 1 (NSD-1) established three NSC sub-groups. The NSC Principals Committee, was composed of the Secretaries of State and Defense, the DCI, the Chairman of the JCS, the Chief of Staff to the President, and the National Security Adviser, who was the chairman. The NSC Deputies Committee, chaired by the Deputy National Security Adviser, was composed of second-ranking officials. There were also a number of NSC Policy Coordinating Committees, chaired by senior officials of the departments most directly concerned with NSC staff members serving as executive secretaries.

The Bush NSC structure most closely resembled that of the Nixon and Ford Administrations in providing for a National Security Adviser chairing most of the key committees. The key differences lay in the personalities involved and the fact that political divisions over foreign policy, while important, lacked some of the emotional heat caused by controversies over Vietnam and Nicaragua. Secretary of State James Baker was a powerful figure in the Administration and a longtime political associate of the President; similarly, Secretary of Defense Dick Cheney himself had White House experience as chief of staff in the Ford Administration and served in a leadership post in the House of Representatives. On occasion, however, Bush did formulate policy within a narrow circle of White House aides.[46]

Evaluation. Whether because of the personalities of NSC principals, the structure of NSC committees or the determination among political opponents to concentrate on the domestic economy, the Bush NSC did not come in for the heavy criticisms that were levied against most of its predecessors. Most observers would probably judge that the Bush Administration created a reasonably effective policymaking machinery and avoided the mistakes of some of its predecessors. Arguably, a standard NSC organization had been created. The Administration successfully addressed most issues that resulted from the breakup of the Soviet Union and the unification of Germany along with the conduct of Desert Storm.

The Clinton NSC, 1993-2001

President Clinton came into office with a determination to focus on domestic issues. His Administration sought to emphasize connections between international concerns and the domestic economy in such areas as trade, banking, and environmental standards. Anthony Lake, who had resigned in protest from the NSC Staff in the Nixon Administration and later served in the State Department in the Carter Administration, was appointed National Security Adviser, and continued in office until he resigned in March 1997. Lake's deputy, Samuel R. Berger, succeeded him, remaining until the end of the Clinton Administration.

With the end of the Cold War, it was widely acknowledged that there was a need for closer integration of national security policy and international economic policy. A major Clinton Administration initiative was the establishment of a National Economic Council (NEC) to coordinate international economic policy which, many observers believed, had usually received short shrift from NSC staffs focused narrowly on diplomatic and security issues. The NEC, initially headed by Robert Rubin who would subsequently become Treasury Secretary in 1995, was charged with coordinating closely with the NSC. To facilitate coordination some NEC staff were "double-hatted" as NSC officials. The close relationship has been credited with enhancing policy coordination at senior White House levels, although, according to some observers, the original promise was not realized as many aspects of international economic and trade policies became parts of major political disputes such as the North American Free Trade Agreement and most-favored-nation status for China.[47]

Some observers would have preferred to include a stronger international economic component within the NSC itself, but others have raised strong objections to such an approach on the grounds that national security policymaking, in significant measure the province of diplomats and military officers, is not as closely related to domestic political concerns as international economic policy. Proponents of the latter view argue that economic issues inevitably involve concerns of various domestic groups and the NSC is ill-suited to integrate them into its policymaking processes.

Presidential Decision Directive (PDD) 2, Organization of the National Security Council, issued on January 20, 1993, expanded the NSC to include, in addition to statutory members and advisers, the Secretary of the Treasury, the U.S. Representative to the United Nations, the Assistant to the President for Economic Policy, and the Chief of Staff to the President. The Attorney General attended relevant meetings including those that discuss covert actions. The National Security Adviser determined the agenda of NSC meetings and ensured the preparation of necessary papers.

The Clinton NSC continued the practice of designating the National Security Adviser as chairman of the Principals Committee of Cabinet-level officers. At a lower level, a Deputies Committee was chaired by the Deputy Assistant to the President for National Security Affairs and included representatives of the key Cabinet departments (as well as the Assistant to the Vice President for National Security Affairs). The Deputies Committee was also responsible for day-to-day crisis management.

In addition, provision was made for a system of Interagency Working Groups, (IWG) some permanent, some *ad hoc*, to be established by the direction of the Deputies Committee and chaired by representatives of the relevant departments, the NEC or the NSC staff. The IWGs convened on a regular basis to review and coordinate the implementation of Presidential decisions in their policy areas.

Evaluation. In general, the Clinton NSC did not see the internecine bureaucratic warfare that had surfaced in earlier administrations. PDD 2 provided for a strong NSC staff. Lake, in his writings on national security policymaking prior to becoming National Security Adviser, reflected a keen appreciation of the disadvantages of bureaucratic infighting. He subsequently recalled that when he came into office, "My model for a national security adviser was that of the behind-the-scenes consensus builder who helped present the communal views of senior advisers to the President." After some months, nonetheless, Lake

> decided to change my approach. I would stay behind the scenes.... And I would do my best always to try to achieve consensus and to make sure that my colleagues' views always had a fair hearing with the President. But I would be less hesitant in voicing my own views when they differed from those of my colleagues, even if it prevented consensus or put me more at odds with them–whether on NATO enlargement, Bosnia, Haiti, or other issues.[48]

In 1999, the Clinton NSC staff played an important and influential role in shaping policy regarding Kosovo. Carefully attuned to shifts in U.S. public opinion, Berger, who succeeded Lake as National Security Adviser in March 1997, reportedly focused on the political dimension of policymaking and sought to avoid options that might lead to paralyzing debate in this country or other NATO states. He is reported to have helped the Administration steer a middle course between those who recommended a ground campaign against Serbia and those more ready to compromise with the Yugoslav leadership and, as a result, the Administration maintained a strong sense of unity throughout the Kosovo campaign. One press account suggested that "What may be Berger's distinctive accomplishment is to have put himself so preeminently at the center of decision-making while minimizing the historic antagonisms between national security advisers and secretaries of state and defense."[49]

The George W. Bush NSC, 2001-2009

In February 2001, President George W. Bush issued National Security Presidential Directive-1, Organization of the National Security Council System." The NSPD indicated that the NSC system was to advise and assist the President and "coordinate executive departments and agencies in the effective development and implementation" of national security policies. Among the statutory and other officials to be invited to attend NSC meetings, the Attorney General will be asked to attend meeting pertaining to his responsibilities, both matters within the Justice Department's jurisdiction and those matters arising under the Attorney General's responsibilities in accordance with 28 USC 511 to give advice and opinion on questions of law. The National Security Adviser was charged with determining the agenda, ensuring necessary papers are prepared and recording NSC actions and presidential decisions.

As has been the custom, the Principals Committee of the NSC consists of relevant department heads and relevant advisory officials, and is chaired by the National Security Adviser. When economic issues are on the agenda the National Security Adviser and the Assistant to the President for Economic Policy are to work in concert. The NSC Deputies Committees will be composed of deputy department heads, advisory officials and is chaired by the Deputy National Security Adviser. Lower-level coordination is effected by Policy Coordinating Committees which are to be chaired by appointees of the Secretary of State, another Cabinet level official or the National Security Adviser.

Subsequent to 9/11, the Intelligence Reform and Terrorism Prevention Act (P.L. 108-458) abolished the position of Director of Central Intelligence and established a new position of Director of National Intelligence (DNI) with enhanced authorities over the entire Intelligence Community. The DNI replaced the DCI in NSC-level deliberations.

Several accounts have described the key role of the NSC in undertaking a review of U.S. options in Iraq in late 2006 that resulted in the changes in tactics and force levels that have come to be known as the Surge. Although senior officials in DOD and the State Department were known to be skeptical of increasing troop levels, NSC staffers are reported to have argued that increased numbers of U.S. forces could provide the security to the Iraqi population that would encourage political stabilization. According to these reports, in the end President Bush adopted this approach.[50]

Evaluation. Although some observers have argued that Condoleezza Rice, as National Security Adviser in President Bush's first term, allowed the Defense Department to dominate policymaking, especially in regard to Iraq, most acknowledged that she had a good working relationship with the President and was an effective public spokesman for the Administration. Hadley made fewer public appearances but emphasized the importance of the NSC staff monitoring the implementation of NSC decisions. As noted above, he is also credited with organizing a review of Iraq policy that resulted in major changes.

The Obama NSC, 2009-

President Obama has designated retired Marine General James L. Jones to serve as National Security Adviser. Jones had previously served as Marine Corps Commandant and as NATO's Supreme Allied Commander. On February 13, 2009, the President signed Presidential Policy Directive-1, *Organization of the National Security Council System.* The directive lists those who will participate in NSC deliberations, including the Attorney General, the Secretary of Homeland Security, the U.S. Representative to the U.N., and the Counsel to the President. It also makes reference to officials who will be specifically invited to sessions dealing with international economic affairs, homeland security, counterterrorism, and science and technology issues. It describes the membership and duties of the Principals and Deputies Committees, which are to be chaired by the National Security Adviser and the Deputy National Security Adviser, respectively. The Principals Committee will be the "senior interagency forum for consideration of policy issues affecting national security" while the Deputies Committee will "review and monitor the work of the NSC interagency process" and "shall be responsible for day-to-day crisis management." The use of the term "monitor" may indicate a determination to enhance the NSC's ability to oversee implementation of presidential decisions on national security issues. Further management of the development and implementation of national security policies will be overseen by interagency policy committees that shall be established to address specific issues.

In May 2009 the Administration announced its intention to integrate the staffs of the National Security Council with the Homeland Security Council into a single National Security Staff, with the goal of ending "the artificial divide between White House staff who have been dealing with national security and homeland security issues.".[51] The position of Assistant to the President for Homeland Security, currently filled by John Brennan, will be retained "with direct and immediate access" to the President, but the incumbent would organizationally report to the National Security Advisor." It is anticipated that the changes will be formalized in a new Presidential Policy Directive.

Evaluation. In the initial months of the Obama Administration, several steps have been taken to modify and enhance the role of the National Security Council. The integration of NSC and Homeland Security Council staffs may work to overcome the intelligence and law enforcement divide that many observers believe existed prior to 9/11. It may also facilitate closer cooperation of Federal agencies and state, local, and tribal entities in dealing with homeland security issues. These relationships are, however, complex and derive from separate statutory missions; observers suggest that establishing new organizational entities can affect, but not determine the ability of different agencies to share information and cooperate on operational planning and programs. The relationships among the relevant senior officials and the role of the President will remain crucial.

Overview of Current NSC Functions

Largely because of the major influence in policymaking exerted by Kissinger and Brzezinski, the position of National Security Adviser has emerged as a central one. Brzezinski was even accorded Cabinet status—the only National Security Adviser to be thus designated.[52] Some observers over the years have argued that the position should be subject to Senate confirmation and that the National Security Adviser should be available to testify before congressional committees as are officials from other Government departments and agencies.[53] Others argue that a President is entitled to confidential advice from his immediate staff. They further suggest that making the position subject to confirmation would create confusion in the eyes of foreign observers as to which U.S. officials speak authoritatively on national security policy. This latter argument is arguably undercut, however, by the practice of recent National Security Advisers of appearing on television news programs.

National Security Advisers have come from various professions; not all have had extensive experience in foreign and defense policy. The report of the Committees Investigating the Iran-Contra Affair recommended that the National Security Adviser not be an active military officer,[54] although no rationale was given for this recommendation.

A substantial number of NSC staff members over the years have been career military or civil servants with backgrounds in foreign policy and defense issues. A considerable number have been detailed to the NSC staff from various federal agencies, which continue to pay their salaries. This practice has been occasionally criticized as allowing the expansion of the White House staff beyond congressional authorization; nonetheless, the practice has continued with annual reports of the number of personnel involved being made to Appropriations Committees.

Beginning with the Kennedy Administration, a concerted effort was made to bring outside experts into the NSC staff in order to inject fresh perspectives and new ideas into the policymaking process. This effort has been continued to varying extents by successive Administrations. Henry Kissinger made a particular effort to hire academic experts, although some would eventually resign and become bitter critics. The Reagan NSC was occasionally criticized for filling NSC staff positions with political activists. Most of the NSC staff positions in the George H.W. Bush Administration were filled with Government officials. Anthony Lake, President Clinton's first National Security Adviser, argues that the NSC staff

> should be made up of as many career officials as possible, with as much carryover between administrations as can be managed. Its experts should be good (but not necessarily gray) bureaucrats who know how to get things done and how to fight for their views, and who are serving the national interest more than the political interests of their President.

He cautioned that:

> a political appointee whose main credential is work on national security issues in political campaigns will have learned to think about national security issues in a partisan context. The effect of his or her advice is likely to be to lengthen the period of time during which a President, at the outset of a term, tries to make policy on the basis of campaign rhetoric rather than international reality.[55]

NSC Executive and Congressional Liaison

The very composition of the NSC, its statutory members, and those who attend meetings on occasion serve to identify those agencies and departments with which the NSC has a regular working relationship. These are the Departments of State and Defense (both the civilian and military staffs), the CIA, the Treasury Department, the Council of Economic Advisers, and a number of other departments as needed. The Director of National Intelligence, who is under the NSC, is responsible for coordinating the nation's foreign intelligence effort. His regular contacts include the CIA, as well as the Defense Intelligence Agency (DIA), the National Security Agency (NSA), the State

Department's Bureau of Intelligence and Research (INR), and other elements of the intelligence community. However, these groups are not represented individually in the current NSC structure.

As part of the Executive Office of the President, the NSC does not have the same regular relationship with Congress and its committees that the member departments and agencies have. Most briefings on intelligence matters are undertaken by the CIA and DIA or by the Director of National Intelligence; information on diplomatic and military matters comes primarily from the Departments of State and Defense. As noted above, the President's Assistant for National Security Affairs is not subject to confirmation by the Senate.

Over the years there have been a considerable number of congressional hearings and reports relating to the NSC. However, many have had to do with topics peculiar to a given period: wiretaps against NSC staff members allegedly ordered by Dr. Kissinger, the unauthorized transfer of NSC documents to officials in the Joint Chiefs of Staff, and Watergate. Annual hearings are held concerning the NSC budget, and there have been occasional hearings concerning NSC organization and procedures. Very few of these hearings and reports have served as briefings for Congress on current issues which the NSC might have been considering. NSC appropriations are handled by the Subcommittees on Financial Services and General Government of the House and Senate Appropriations Committees.

As has been noted, Congress's role in NSC matters and its relationship with the NSC are limited. The Senate does not approve the appointment of the National Security Adviser, although it does confirm statutory NSC members Congress does have authority over the designation of those positions that are to have statutory NSC membership, as well as budgetary authority over the NSC. In 2007, as part of the Energy Independence and Security Act of 2007 (P.L. 110-140, section 932) Congress added the Secretary of Energy to the NSC. However, Congress has little direct say in matters of NSC organization, procedure, role, or influence, although a number of hearings on these topics have been held.

The NSC is not a primary and regular source of national security information for Congress. National security information is for the most part provided by those departments and agencies that are represented on the NSC. The NSC, as a corporate entity, rarely testifies before or briefs Congress on substantive questions, although in some Administrations informal briefings have been provided.

The NSC is an organ devoted to the workings of the executive branch in the broad area of national security. Its role is basically that of policy analysis and coordination and, as such, it has been subject to limited oversight and legislative control by Congress. Both in its staff organization and functioning, the NSC is extremely responsive to the preferences and working methods of each President and Administration. It would be difficult to design a uniform NSC structure that would meet the requirements of chief executives who represent a wide range of backgrounds, work styles, and policy agendas although some observers believe that the general pattern established in the final years of the Reagan Administration and followed by successive Presidents is likely to endure. There is unlikely to be a desire to drastically reduce the role of the NSC staff and most observers suggest that elevating the policymaking role of the National Security Adviser at the expense of the Secretary of State leaves Presidents subject to strong criticism.

The NSC and International Economic Issues

The NSC has traditionally focused on foreign and defense policy issues. In the aftermath of the end of the Cold War, many observers argue that the major national security concerns of the United States may no longer be centered on traditional diplomatic and military issues. They suggest, further, that international economic, banking, environmental, and health issues, among others, will be increasingly important to the country's national security. These types of concerns, however, have not been regularly part of the NSC's primary areas of responsibility. The heads of federal agencies most directly concerned with such issues have not been members of the NSC.[56]

In the 1970's, Maxwell Taylor, who President Kennedy had appointed Chairman of the JCS, argued that a National Policy Council should replace the NSC and concern itself with broad areas of international and domestic policy.[57] William Hyland, an NSC official in the Reagan Administration, argued in 1980 that

> ... a bad defect in the [NSC] system is that it does not have any way of addressing international economic problems. The big economic agencies are Treasury, to some extent OMB, the Council of Economic Advisers, Commerce, Labor, and Agriculture. They are not in the NSC system, but obviously energy problems, trade, and arms sales are foreign policy issues. Every Administration tries to drag them in, usually by means of some kind of

> a subcommittee or a separate committee. The committee eventually runs up against some other committee. There is friction, and policies are made on a very ad hoc basis by the principal cabinet officers.[58]

In early 1992, Professor Ernest May of Harvard University testified to the Senate Select Committee on Intelligence:

> In the early 1980s, the greatest foreign threat was default by Mexico and Brazil. That could have brought down the American banking system. Despite good CIA analysis and energetic efforts by some NSC staffers, the question did not get on the NSC agenda for more than two years. And then, the policy issues did not get discussed. The agencies concerned with money and banking had no natural connection with either the NSC or the intelligence community. We have no reason to suppose that agencies concerned with the new [post Cold War] policy issues will be any more receptive.[59]

In the George H.W. Bush Administration, there remained a strong conviction that defense and foreign policy issues would remain vital and somewhat separate from other interests and that the NSC was the proper forum for them to be addressed. Before he became President Bush's National Security Adviser, Brent Scowcroft stated at a forum on national security policy organization:

> First of all, if there is a consensus ... that the NSC net ought to be spread ever wider, I am not a part of it. There are many things that the NSC system can do better, and it has enough on its plate now. I would not look toward its spreading its net wider.[60]

As noted above, the Clinton Administration implemented its determination to coordinate foreign and domestic economic policies more closely. The National Economic Council, established by Executive Order 12835 on January 25, 1993, was designed to "coordinate the economic policymaking process with respect to domestic and international economic issues." Close linkage with the NSC were to be achieved by having the Assistant to the President for Economic Policy also sit on the NSC, supplemented by assigning staff to support both councils. The goal was to ensure that the economic dimensions of national security policy would be properly weighed in the White House decision-making process. Observers consider that cooperation between the NSC and the NEC was productive and contributed to the enhancement of both national security and economic policymaking although one senior NSC official has noted that efforts to deal with the 1997 Asian financial crisis were initially coordinated by U.S. international economic policymakers with little input from national security and foreign policy agencies.[61]

The Growing Importance of Law Enforcement Issues

The post-Cold War era has seen a much closer relationship between traditional national security concerns with international issues that have a significant law enforcement component such as terrorism and narcotics smuggling.[62] The increasing intermingling of national security and law enforcement issues could cause major difficulties for the NSC staff and the National Security Adviser who is not a law enforcement official. The Justice Department will inevitably view with concern any incursion into what is regarded as the Attorney General's constitutional responsibilities. The NSC also coordinates with the Office of Drug Control Policy whose responsibilities also encompass both law enforcement and foreign policy considerations.

In dealing with international terrorism or narcotics production and transport from foreign countries, however, diplomatic and national security issues are often involved. Apprehending a terrorist group may require cooperation from a foreign government that has its own interests and concerns. Narcotics production may be entwined in the social and economic fabric of a foreign country to an extent that precludes the country from providing the sort of cooperation that would be expected from a major ally. During the Clinton Administration, the Attorney General's representatives have been included in NSC staff deliberations when law enforcement concerns were involved. Nonetheless, observers note public disagreements between Justice Department and State Department, for instance, regarding cooperation (or the lack thereof) from Saudi Arabia or Yemen. Clearly, the President has constitutional responsibilities for both national security and law enforcement, but the status of any other official to make necessary trade-offs is unclear. Observers suggest that in some future cases the need to establish a single U.S. position may require different ways of integrating national security and law enforcement concerns. The Obama Administration took a significant step in this direction by establishing a single National Security Staff composed of the staffs of both the National Security Council and the Homeland Security Council.

Today's international terrorist threat can encompass not only physical attacks on U.S. physical structures such as the World Trade Center, but also cyber-attacks on critical infrastructures, the computerized communications and data storage systems on which U.S. society has become reliant. Since such systems are in most cases owned and operated by corporations and other commercial entities, the role of the NSC is necessarily constrained. Much depends on law enforcement as well as voluntary cooperation by the private sector. The Clinton Administration created the position of National Coordinator for Security, Infrastructure Protection, and Counterterrorism who reported to the President through the National Security Adviser.[63] The Intelligence Reform Act of 2004, however, established the National Counterterrorism Center outside the NSC structure.

In dealing with policies related to the protection of critical infrastructures, the National Security Adviser will have an important role, but one inherently different from the traditional responsibilities of the office.[64] The position could involve in coordination of responses to threats both in the U.S. and from abroad and among the federal government, the states, and the private sector. It is clear to all observers that such coordination involves much uncharted territory, including a concern by some that the National Security Adviser might become overly and inappropriately involved in law enforcement matters.

The Role of the National Security Adviser

The NSC was created by statute, and its membership has been designated and can be changed by statute. The NSC has also been subject to statutorily approved reorganization processes within the executive branch, as when it was placed in the Executive Office by a Reorganization Plan in August 1949. Nonetheless, the NSC has been consistently regarded as a presidential entity with which Congress is rarely involved. The internal organization and roles of the NSC have been changed by Presidents and by National Security Advisers in response to their preferences and these changes have not usually been subject to congressional scrutiny.

The role of the National Security Adviser has, however, become so well established in recent years that Congress has been increasingly prepared to grant the incumbent significant statutory responsibilities. The Foreign Intelligence Surveillance Act and other legislation provides for statutory roles for the National Security Adviser.[65] Executive Orders provide other formal responsibilities. The position has become institutionalized and the exercise of its functions has remained an integral part of the conduct of national security policy in all recent administrations.

Some observers believe that these established duties which extend beyond the offering of advice and counsel to the President will inevitably lead to a determination to include the appointment of a National Security Adviser among those requiring the advice and consent of the Senate. Advice and consent by the Senate is seen as providing a role for the legislative branch in the appointment of one of the most important officials in the federal government. Another cited advantage of this proposal would be the increased order, regularity, and formalization that are involved in making appointments that are sent to the Senate. Proponents argue that this would ultimately provide greater accountability for NSC influence and decisions. Opponents on the other hand, might point to the danger of unnecessary rigidity and stratification of organization and the potential that appointments might be excessively influenced by political considerations. There is also a potential that the NSC staff might become irrelevant if it loses the trust of a future President or if its procedures become so formalized as to stultify policymaking. Should the Adviser be subject to Senate confirmation, it is argued that an important prerogative of the President to choose his immediate staff would be compromised. In addition, the incumbent could be required or expected to make routine appearances before congressional oversight committees, arguably undermining the primary purpose of the National Security Adviser which is to provide the President with candid advice on a wide range of issues, often on an informal and confidential basis.

One historian has summed up the role of the National Security Adviser:

> The entire national security system must have confidence that the [National Security Adviser] will present alternate views fairly and will not take advantage of propinquity in the coordination of papers and positions. He must be able to present bad news to the president and to sniff out and squelch misbehavior before it becomes a problem. He must be scrupulously honest in presenting presidential decisions and in monitoring the implementation process. Perhaps most important, he must impart the same sense of ethical behavior to the Staff he leads.[66]

In a recent assessment, two informed observers listed tasks for which the Adviser and staff uniquely are responsible:

- Staffing the president's daily foreign policy activity: his communications with foreign leaders and the preparation and conduct of his trips overseas;
- Managing the process of making decisions on major foreign and national security issues;
- Driving the policymaking process to make real choices, in a timely manner;
- Overseeing the full implementation of the decision the president has made.[67]

The increasing difficulties in separating national security issues from some law enforcement and international economic concerns has led some observers to urge that the lines separating various international staffs at the White House be erased and that a more comprehensive policymaking entity be created. It is argued that such reforms could most effectively be accomplished without legislation.[68] The Project for National Security Reform (PNSR) a nonpartisan task force that has studied the structure of national policymaking has made a number of far-reaching proposals for expanding the role of a national security director and combining the National Security Council and the Homeland Security Council.[69] As noted above, President Obama put this latter recommendation into practice in May 2009. Such initiatives raise complex questions, including the role of congressional oversight. Whereas Congress has traditionally deferred to White House leadership in national security matters, to a far greater extent than in international economic affairs, there might be serious questions about taking formal steps to place resolution of a wide range of international policies, including economic and law enforcement issues, in the hands of officials who receive little congressional oversight. It is likely, in any event, that Congress will continue to monitor the functioning of the staff and the Adviser in the context of U.S. policymaking in a changing international environment.

SELECTED BIBLIOGRAPHY

The most comprehensive source concerning the genesis and development of the NSC through 1960 is contained in a collection of hearings, studies, reports and recommendations complied by Senator Henry M. Jackson and published as U.S. Congress. Senate. Committee on Government Operations.

Subcommittee on National Policy Machinery. *Organizing for National Security*, 3 Vols. 86th and 87th Congress. Washington, Government Printing Office. [1961]. Presidential memoirs are also valuable.

Other useful sources are:

[1] Anderson, Dillon. (1966)."The President and National Security." *Atlantic Monthly*, January.

[2] Bock, Joseph G. (1986). "The National Security Assistant and the White House Staff: National Security Policy Decisionmaking and Domestic Political Considerations, 1947-1984." *Presidential Studies Quarterly*, Spring. 258-279.

[3] Bowie, Robert R. & Richard H. (1998). Immerman. *Waging Peace: How Eisenhower Shaped an Enduring Cold War Strategy*. New York: Oxford University Press.

[4] Brown, Cody M. (2008). *The National Security Council: A Legal History of the President's Most Powerful Advisers*. Washington: Project for National Security Reform.

[5] Brzezinski, Zbigniew. (1983). *Power and Principle: Memoirs of the National Security Adviser, 1977-1981*. New York: Farrar, Straus, Giroux.

[6] ——"The NSC's Midlife Crisis". *Foreign Policy*, Winter 1987-1988. 80-99.

[7] Bock, Joseph G. & Clarke, Duncan L. (1986). "The National Security Assistant and the White House Staff: National Security Policy Decisionmaking and Domestic Political Considerations, 1947-1984." *Political Science Quarterly*, Spring 1986. 258-279.

[8] Brown, Cody M. (2008). *The National Security Council: A Legal History of the President's Most Powerful Advisers*. Washington: Project for National Security Reform.

[9] Bumiller, Elisabeth. (2007).*Condoleezza Rice: An American Life: A Biography*. New York: Random House.

[10] Burke, John P. (2009). *Honest Broker: the National Security Advisor and Presidential Decision Making*. College Station, TX: Texas A & M University Press.

[11] Caraley, Demetrios. (1966). *The Politics of Military Unification*. New York: Columbia University Press.

[12] Clark, Keith C. & Laurence S. Legere, eds. (1966). "The President and the Management of National Security". Report for the Institute for Defense Analyses. New York: Praeger. See especially 55-114.

[13] Commission on the Organization of the Executive Branch of the Government. (1949). *National Security Organization*. Washington: Government Printing Office.
[14] Commission on the Organization of the Government for the Conduct of Foreign Policy. (1975). *Report*. Washington: Government Printing Office.
[15] Cutler, Robert. (1966). *No Time for Rest*. Boston: Little, Brown.
[16] ——"The Development of the National Security Council". *Foreign Affairs*, April 1956. 441-58.
[17] Daalder, Ivo H. & Destler, I.M. (2009). *In the Shadow of the Oval Office: Profiles of the National Security Advisers and the Presidents They Served—from JFK to George W. Bush*. New York: Simon and Schuster.
[18] Destler, I.M. (1971-72). "Can One Man Do?" *Foreign Policy*, Winter. 28-40.
[19] ——"National Security Advice to U.S. Presidents: Some Lessons from Thirty Years." *World Politics*, January 1977. 143-76.
[20] ——*Presidents, Bureaucrats and Foreign Policy: The Politics of Organization*. Princeton: Princeton University Press. (1972).
[21] ——Lake, Anthony, & Gelb, Leslie H. *Our Own Worst Enemy: The Unmaking of American Foreign Policy*. New York: Simon and Schuster. 1984.
[22] Falk, Stanley L. & Theodore W. Bauer. (1972). "The National Security Structure." Washington: Industrial College of the Armed Forces.
[23] Greenstein, Fred I. & Richard H. Immerman. (2000). "Effective National Security Advising: Recovering the Eisenhower Legacy." *Political Science Quarterly*, Fall. 335-345.
[24] Haig, Alexander M., Jr. (1984).*Caveat: Realism, Reagan, and Foreign Policy*. New York: Macmillan.
[25] Hammond, Paul Y. (1960). "The National Security Council as a Device for Interdepartmental Coordination: An Interpretation and Appraisal". *American Political Science Review*, December. 899-910.
[26] Humphrey, David C. (1994). "NSC Meetings during the Johnson Presidency". *Diplomatic History*, Winter. 29-45.
[27] Hunter, Robert E. (1988). *Organizing for National Security*. Washington: Center for Strategic and International Studies.
[28] Inderfurth, Karl F. & Johnson, Loch K., eds. (1988).*Decisions of the Highest Order: Perspectives of the National Security Council*. Pacific Grove, CA: Brooks/Cole Publishing Co.

[29] Johnson, Robert H. (1969). "The National Security Council: The Relevance of its Past to its Future". *Orbis*, Fall. 709-35.
[30] Kissinger, Henry A. (1979). *White House Years*. Boston: Little, Brown.
[31] —— *Years of Renewal*. New York: Simon & Schuster. 1999.
[32] —— *Years of Upheaval*. Boston: Little, Brown. 1982.
[33] Kolodziej, Edward A. (1969). "The National Security Council: Innovations and Implications". *Public Administration Review*. November/December. 573-85.
[34] Korb, Lawrence J. & Hahn, Keith D. eds. (1981). *National Security Policy Organization in Perspective*. Washington: American Enterprise Institute.
[35] Laird, Melvin R. (1987). *Beyond the Tower Commission*. Washington: American Enterprise Institute.
[36] Lake, Anthony. (2000). *6 Nightmares*. Boston: Little, Brown.
[37] —— *Somoza Failing*. Boston: Houghton Mifflin. 1989.
[38] Lay, James S., Jr. (1952). "National Security Council's Role in the U.S. Security and Peace Program". *World Affairs*, Summer. 33-63.
[39] Leacacos, John P. (1971-72). "Kissinger's Apparat". *Foreign Policy*, Winter. 3-27.
[40] Lord, Carnes. (2000). "NSC Reform for the Post-Cold War Era," *Orbis*, Summer . 433-450.
[41] McFarlane, Robert C. (1988). "Effective Strategic Policy." *Foreign Affairs*, Fall. 33-48.
[42] ——with Richard Saunders and Thomas C. Shull. "The National Security Council: Organization for Policy Making." *Proceedings of the Center for the Study of the Presidency*. Vol. 5. 1984. 261-273.
[43] ——Head, Richard G., and Frisco W. Short. *Crisis Resolution, Presidential Decision Making in the Mayaguez and Korean Confrontations*. Boulder, CO: Westview Press. 1978.
[44] Menges, Constantine C. *Inside the National Security Council: The True Story of the Making and Unmaking of Reagan's Foreign Policy*. New York: Simon and Schuster. 1988.
[45] Mulcahy, Kevin V. & Crabb, Cecil V. (1991). "Presidential Management of National Security Policy Making, 1947-1987". In *The Managerial Presidency*, ed. by James P. Pfiffner. Pacific Grove, CA: Brooks/Cole. 250-264.
[46] Nelson, Anna Kasten. (1985). "President Truman and the Evolution of the National Security Council." *Journal of American History*, September. 360-378.

[47] ——"The 'Top of the Policy Hill': President Eisenhower and the National Security Council." *Diplomatic History*, Fall 1983. 307-326.

[48] Nixon, Richard M. *U.S. Foreign Policy for the 1970's: A New Strategy for Peace*. Washington: Government Printing Office. (1970).

[49] ——*U.S. Foreign Policy for the 1970's: The Emerging Structure of Peace*. Washington: Government Printing Office. (1971).

[50] ——*U.S. Foreign Policy for the 1970's: Building for Peace*. Washington: Government Printing Office. (1972).

[51] Powell, Colin L. (1989). "The NSC System in the Last Two Years of the Reagan Administration". *The Presidency in Transition*. Ed. by James P. Pfiffner and R. Gordon Hoxie. New York: Center for the Study of the Presidency, 204-218.

[52] Prados, John. (1991). *Keepers of the Keys: A History of the National Security Council from Truman to Bush*. New York: William Morrow.

[53] Rothkopf, David. (2004). *Running the World: the Inside Story of the National Security Council and the Architects of American Power*. New York: Public Affairs.

[54] Rostow, Walt Whitman. (1972). *The Diffusion of Power: An Essay in Recent History*. New York: Macmillan.

[55] Sander, Alfred D. (1972). "Truman and the National Security Council, 1945-1947". *Journal of American History*, September. 369-388.

[56] Schlesinger, Arthur, Jr. (2000). "Effective National Security Advising: A Most Dubious Precedent." *Political Science Quarterly*, Fall. 347-351.

[57] Shoemaker, Christopher C. (1991) .*The NSC Staff: Counseling the Council*. Boulder, CO: Westview Press.

[58] Souers, Sidney W. (1949). "Policy Formulation for National Security". *American Political Science Review*, June 1949. 534-43.

[59] Steiner, Barry H. (1977). "Policy Organization in American Security Affairs: An Assessment". *Public Administration Review*, July/August. 357-67.

[60] Thayer, Frederick C. (1971). "Presidential Policy Process and 'New Administration': A Search for Revised Paradigms". *Public Administration Review*, September/October. 552-61.

[61] U.S. Congress. Senate. (1969).Committee on Government Operations. Subcommittee on National Security and International Operations. *The National Security Council: New role and structure*. 91st Congress, 1st session. Washington: Government Printing Office.

[62] ——Select Committee on Secret Military Assistance to Iran and the Nicaraguan Opposition. House of Representatives. Select Committee to

Investigate Covert Arms Transactions with Iran. 100th Congress, 1st session. *Report of the Congressional Committees Investigating the Iran-Contra Affair with Supplemental, Minority, and Additional Views*, Senate Report 100-216/House Report 100-433. Washington: Government Printing Office. 1987.

[63] ——Committee on Government Operations. Subcommittee on National Security and International Operations. *The National Security Council: Comment by Henry Kissinger*. March 3, 1970. 91st Congress, 2d Session. Washington: Government Printing Office. 1970.

[64] U.S. Department of State. (1969). "The National Security System: Responsibilities of the Department of State." *Department of State Bulletin*, February 24, 163-66.

[65] U.S. President's Special Review Board. (1987). *Report of the President's Special Review Board*. Washington: Government Printing Office.

[66] West, Bing. (2008). *The Strongest Tribe: War, Politics and the Endgame in Iraq*. New York. Random House.

[67] Yost, Charles W. (1971). "The Instruments of American Foreign Policy," *Foreign Affairs*, October. 59-68.

[68] Zegart, Amy B. (1999). *Flawed by Design: the Evolution of the CIA, JCS, and NSC*. Stanford, CA: Stanford University Press.

Note: Many of the above entries contain numerous footnotes that identify a wealth of primary and secondary sources too numerous to include here. Of special interest are the oral interviews of former NSC staff personnel conducted from 1998 to 2000 as part of the National Security Council Project undertaken by the Center for International and Security Studies at Maryland and the Brookings Institution; transcripts are available at http://www.cissm.umd.edu/projects/nsc.php. Also useful is the transcript of "A Forum on the Role of the National Security Adviser," cosponsored by the Woodrow Wilson International Center for Scholars and the James A. Baker III Institute for Public Policy of Rice University, available at *http://wwics.si.edu* /news/docs/nsa.pdf.

Appendix A. National Security Advisers

Table A-1. National Security Advisers, 1953-2009

Robert Cutler	March 23, 1953	April 2, 1955
Dillon Anderson	April 2, 1955	September 1, 1956
Robert Cutler	January 7, 1957	June 24, 1958
Gordon Gray	June 24, 1958	January 13, 1961
McGeorge Bundy	January 20, 1961	February 28, 1966
Walt W. Rostow	April 1, 1966	January 20, 1969
Henry A. Kissinger	January 20, 1969	November 3, 1975
Brent Scowcroft	November 3, 1975	January 20, 1977
Zbigniew Brzezinski	January 20, 1977	January 21, 1981
Richard V. Allen	January 21, 1981	January 4, 1982
William P. Clark	January 4, 1982	October 17, 1983
Robert C. McFarlane	October 17, 1983	December 4, 1985
John M. Poindexter	December 4, 1985	November 25, 1986
Frank C. Carlucci	December 2, 1986	November 23, 1987
Colin L. Powell	November 23, 1987	January 20, 1989
Brent Scowcroft	January 20, 1989	January 20, 1993
W. Anthony Lake	January 20, 1993	March 14, 1997
Samuel R. Berger	March 14, 1997	January 20, 2001
Condoleezza Rice	January 22, 2001	January 25, 2005
Stephen Hadley	January 26, 2005	January 20, 2009
James L. Jones	January 20, 2009	Present

End Notes

[1] Paul Y. Hammond, "The National Security Council as a Device for Interdepartmental Coordination: An Interpretation and Appraisal," *American Political Science Review*, December, 1960, p. 899; U.S. Bureau of the Budget, *The United States at War* (Washington: Government Printing Office, 1946), p. 2.

[2] Mark Skinner Watson, *Chief of Staff: Prewar Plans and Preparations* (Washington: Office of the Chief of Military History, 1950), pp. 89-91, 93-94.

[3] R. Elberton Smith, *The Army and Economic Mobilization* (Washington: Office of the Chief of Military History, 1959), pp. 103-04, 109-10; Bureau of the Budget, *The United States at War*, pp. 22-25, 44, 50-51.

[4] Ray S. Cline, *Washington Command Post: The Operations Division* (Washington: Office of the Chief of Military History, 1951), pp. 326-27; John Lewis Gaddis, *The United States and the Origins of the Cold War* (New York, Columbia University Press, 1972), p. 126; U.S.

Department of State, *Foreign Relations of the United States*, 1944, v. I: *General* (Washington: Government Printing Office, 1966), pp. 1466-70.

[5] One of the best studies on the creation and development of the NSC through the Eisenhower Administration, including hearings, studies, reports, recommendations and articles, can be found in U.S. Congress, Senate, 86th and 87th Congress, Committee on Government Operations, Subcommittee on National Policy Machinery, *Organizing for National Security*, 1961, 3 vols.

[6] Demetrios Caraley, *The Politics of Military Unification* (New York: Columbia University Press, 1966), pp. 23-44; Walter Millis, ed., *The Forrestal Diaries* (New York: The Viking Press, 1951), pp. 62-63.

[7] Caraley, *The Politics of Military Unification*, pp. 40-41; see also Jeffrey M. Dorwart, *Eberstadt and Forrestal: A National Security Partnership, 1909-1949* (College Station, TX: Texas A & M University Press, 1991), especially pp. 90-107.

[8] Ibid., pp. 86-87, 91; Hammond, "The NSC as a Device for Interdepartmental Coordination," pp. 900-01.

[9] Caraley, *Politics of Military Unification*, pp. 136-37; Millis, *Forrestal Diaries*, p. 222.

[10] The congressional debate over the National Security Act is summarized in Caraley, *Politics of Military Unification*, pp. 153-82; on the NSC, see p. 161. Examples of congressional opinion can be found throughout the lengthy debate. Some representative comments can be found in the *Congressional Record*, v. 93, July 7, 1947, p. 8299, and July 9, 1947, pp. 8496-97, 8518, 8520.

[11] 50 USC 402.

[12] More specific information on the history of the transfers of defense mobilization and civil defense authorities may be found in Sections 402 and 404 of *U.S. Code Annotated*, Title 50 (St. Paul, MN: West Publishing Co., 1991).

[13] The Commission on the Organization of the Executive Branch of the Government. *National Security Organization* (Washington: Government Printing Office, 1949), especially pp. 15-16, 74-76.

[14] Walter Millis, *Arms and the State* (New York: The Twentieth Century Fund, 1958), pp. 255, 388.

[15] Anna Kasten Nelson, "President Truman and the Evolution of the National Security Council," *Journal of American History*, September 1985, p. 377.

[16] This position has been a continuing one, although its title has varied over the years (Special Assistant for National Security Affairs, Assistant to the President for National Security Affairs, National Security Adviser). An adviser or an assistant to the President arguably has a position of greater independence from congressional oversight than the incumbent of a position established by statute; see Richard Ehlke, "Congressional Creation of an Office of National Security Adviser to the President," reprinted in U.S. Congress, 96th Congress, 2d session, Senate, Committee on Foreign Relations, *The National Security Adviser: Role and Accountability*, Hearing, April 17, 1980, pp. 133-135. The position of National Security Adviser is to be distinguished from the position of Executive Secretary of the NSC, which was created by statute but has, since the beginning of the Eisenhower Administration, been essentially an administrative and logistical one. National Security Adviser positions are funded not as part of the NSC but as part of the White House Office, reflecting the incumbent's status as that of an adviser to the President.

[17] Frederick C. Thayer, "Presidential Policy Processes and 'New Administration': A Search for Revised Paradigms," *Public Administration Review*, September/October 1971: 554. Robert H. Johnson, "The National Security Council: The Relevance of its Past to Its Future," *Orbis*, v. 13, Fall 1969: 715; Robert Cutler, *No Time for Rest* (Boston: Little, Brown, 1966).

[18] Some Eisenhower-era NSC documents are reprinted in the State Department's *Foreign Relations of the United States* series, especially in volumes dealing with National Security Affairs; microfilm copies of declassified NSC documents have been made available by a commercial publisher, University Publications of America, Inc. NSC documents are usually

highly classified; while some are subsequently declassified and released to the public (although not necessarily in the *Federal Register* or other official publications series), others have been withheld. Some observers have criticized this situation; see Harold C. Relyea, "The Coming of Secret Law," *Government Information Quarterly*, May 1988, pp. 106-112; U.S. General Accounting Office, "The Use of Presidential Directives to Make and Implement U.S. Policy," Report No. GAO/NSIAD-92-72, January 1992.

[19] The hearings and reports of this study are cited in note 5.

[20] See Anna Kasten Nelson, "The 'Top of Policy Hill': President Eisenhower and the National Security Council," *Diplomatic History*, Fall 1983, p. 324; also, Stephen E. Ambrose, *The President* (New York: Simon and Schuster, 1984), pp. 345, 509.

[21] Robert R. Bowie and Richard H. Immerman, *Waging Peace: How Eisenhower Shaped an Enduring Cold War Strategy* New York: Oxford University Press, 1998), p. 258. Recent support for the Eisenhower system is given in Fred I. Greenstein and Richard H. Immerman, "Effective National Security Advising: Recovering the Eisenhower Legacy," *Political Science Quarterly*, Fall 2000; criticism is renewed in Arthur Schlesinger, Jr., "Effective National Security Advising: A Most Dangerous Precedent," *Political Science Quarterly*, Fall 2000.

[22] Thayer, "Presidential Policy Processes and 'New Administration,'" p. 555.

[23] Arthur M. Schlesinger, Jr., *A Thousand Days: John F. Kennedy in the White House* (Boston: Houghton Mifflin, 1965), pp. 406-47; see especially pp. 412-13, 426, 430-32.

[24] Kai Bird, *The Color of Truth: McGeorge Bundy and William Bundy: Brothers in Arms* (New York: Simon & Schuster, 1998), p. 186.

[25] McGeorge Bundy to Henry M. Jackson, September 4, 1961, reprinted in *Organizing for National Security*, I, 1338.

[26] Much of the documentary basis of the Nixon NSC effort is provided in U.S., Department of State, *Foreign Relations of the United States, 1969-1976*, Vol. II, *Organization and Management of U.S. Foreign Policy, 1969-1972* (Washington, DC: Government Printing Office, 2006).

[27] Commission on the Organization of the Government for the Conduct of Foreign Policy, Report (Washington, Government Printing Office, 1975).

[28] Ibid., p. 37.

[29] Veto of a Bill to Amend the National Security Act of 1947, January 1, 1976, *Public Papers of the Presidents: Gerald R. Ford* 1976-1977 Vol. I (Washington: Government Printing Office, 1979), p. 1-2.

[30] For the Carter NSC, see Presidential Directive/NSC-1 and NSC-2, January 20, 1977; Executive Order 12036, January 24, 1978; Zbigniew Brzezinski, *Power and Principle: Memoirs of the National Security Adviser, 1977-1981* (New York: Farrar, Straus, Giroux, 1983).

[31] Presidential Directive/NSC-2, January 20, 1977; see also Statement by White House Press Secretary, January 22, 1977, *Public Papers of the Presidents, Jimmy Carter*, 1977, Vol. I (Washington: Government Printing Office, 1977), p. 8..

[32] Presidential Directive/NSC-2.

[33] Executive Order 12036, January 24, 1978, Section 1-2.

[34] Presidential Directive/NSC-2. January 20, 1977.

[35] Executive Order 12036, January 24, 1978, Section 1-3.

[36] Executive Order 12036, January 24, 1978, Sections 1-302 to 1-304 inclusive.

[37] See Christopher C. Shoemaker, *The NSC Staff: Counseling the Council* (Boulder, CO: Westview Press, 1991), pp. 51-57.

[38] Stansfield Turner, *Terrorism and Democracy* (Boston: Houghton Mifflin, 1991), p. 58.

[39] See *National Security Adviser: Role and Accountability*.

[40] Brzezinski acknowledges but discounts the reports that Mondale had imposed certain staff members on him; see *Power and Principle*, pp. 74-78.

[41] Reprinted in *Public Papers of the Presidents, Ronald Reagan*, 1982, Vol. I (Washington: Government Printing Office, 1983), pp. 18-22.

[42] *Public Papers of the Presidents, Ronald Reagan*, 1981 (Washington: Government Printing Office, 1982), p. 285.

[43] U.S., President's Special Review Board, *Report of the President's Special Review Board* (Washington: Government Printing Office, 1987), pp. V-1-V-2. The Board consisted of former Senator John Tower, former Secretary of State Edmund Muskie and former (and future) National Security Adviser Brent Scowcroft. It is often referred to as the Tower Board.

[44] Ibid., p. V-5.

[45] U.S. Congress. Senate. Select Committee on Secret Military Assistance to Iran and the Nicaraguan Opposition. House of Representatives. Select Committee to Investigate Covert Arms Transactions with Iran. 100th Congress, 1st session. *Report of the Congressional Committees Investigating the Iran-Contra Affair with Supplemental, Minority, and Additional Views*, Senate Report 100-216/House Report 100-433, November 1987, p. 425. The Committee added that, "By statute the NSC was created to provide advice to the President on national security matters. But there is no express statutory prohibition on the NSC engaging in operational intelligence activities." Ibid.

[46] See Robert G. Sutter, "American Policy Toward Beijing, 1989-1990: the Role of President Bush and the White House Staff," *Journal of Northeast Asian Studies*, Winter 1990.

[47] The NEC was established by Executive Order 12835 on January 25, 1993; on the NEC see Kenneth I. Jester and Simon Lazarus, *Making Economic Policy: An Assessment of the National Economic Council* (Washington: Brookings Institution Press, 1997) and I.M. Destler, *The National Economic Council: A Work in Progress* (Washington: Institute for International Economics, 1996).

[48] *6 Nightmares* (Boston: Little, Brown, 2000), pp. 131,131-132.

[49] John F. Harris, "Berger's Caution Has Shaped Role of U.S. In War," *Washington Post*, May 16, 1999, p. A24.

[50] See Linda Robinson, *Tell Me How This Ends: General David Petraeus and the Search for a Way Out of Iraq* (New York: Public Affairs, 2008), pp. 26-36; Bing West, *The Strongest Tribe: War, Politics, and the Endgame in Iraq* (New York: Random House, 2008), pp. 197-207; Ivo H. Daalder and I.M. Destler, *In the Shadow of the Oval Office: Profiles of the National Security Advisers and the Presidents they Served from JFK to George W.* Bush (New York: Simon and Schuster, 2009), pp. 294-297.

[51] Statement by the President on the White House Organization for Homeland Security and Counterterrorism, Office of the Press Secretary, May 26, 2009.

[52] "Cabinet status" is not recognized in law, but is a distinction conferred by the President. See Ronald C. Moe, *The President's Cabinet*, CRS Report No. 86-982GOV, November 6, 1986, p. 2.

[53] There are differing views regarding linkage between Senatorial confirmation and an obligation to testify before congressional committees; see, for instance, Zbigniew Brzezinski, "NSC's Midlife Crisis," *Foreign Policy*, Winter 1987-1988, p. 95; also, the Prepared Statement of Thomas M. Franck printed in *The National Security Adviser: Role and Accountability*, pp. 40-41. For additional background, see CRS Report RL31351, *Presidential Advisers' Testimony Before Congressional Committees: An Overview*, by Harold C. Relyea and Todd B. Tatelman.

[54] Report of the Congressional Committees Investigating the Iran-Contra Affair, p. 426.

[55] *6 Nightmares*, pp. 261-262.

[56] There are other White House-level coordinative bodies, such as the Office of Management and Budget, the Council of Economic Advisers, the Office of the United States Trade Representative, and the Council on Environmental Quality, that do deal with such issues.

[57] Maxwell D. Taylor, *Precarious Security* (New York: W.W. Norton, 1976), pp. 113-116.

[58] Quoted in Lawrence J. Korb and Keith D. Hahn, eds., *National Security Policy Organization in Perspective* (Washington: American Enterprise Institute, 1980), p. 11.

[59] Ernest R. May, Statement for the Senate Select Committee on Intelligence, March 4, 1992; see also May's article, "Intelligence: Backing into the Future," *Foreign Affairs*, Summer 1992, especially pp. 64-66.

[60] Quoted in Korb and Hahn, eds., *National Security Policy Organization in Perspective*, p. 34.

[61] James Steinberg, "Foreign Policy: Time to Regroup," *Washington Post*, January 2, 2001, p. A15.

[62] See CRS Report RL30252, *Intelligence and Law Enforcement: Countering Transnational Threats to the U.S.*, by Richard A. Best Jr..

[63] See CRS Report RL30153, *Critical Infrastructures: Background, Policy, and Implementation*, by John D. Moteff.

[64] It has been noted that the original membership of the National Security Council included officials responsible for mobilization planning, but those offices were subsequently merged into others and are no longer represented on the NSC. The need for national mobilization to sustain a global war effort is not considered a high priority in the post-Cold War world. (See Carnes Lord, "NSC Reform in the Post-Cold War Era," *Orbis*, Summer 2000, pp. 449-500.) The inclusion of such officials did, nonetheless, reflect the determination of the drafters of the National Security Act that the NSC have a wide mandate in protecting the nation's security interests and one that could extend into the private sector.

[65] The Foreign Intelligence Surveillance Act of 1978, as amended, requires that applications for orders for electronic surveillance for foreign intelligence purposes include a certification regarding the need for such surveillance by the Assistant to the President for National Security Affairs (or someone else designated by the President)(50 USC 1804 (a)(7)); a similar requirement exists for applications for physical searches (50 USC 1823(a)(7)). The Assistant to the President for National Security Affairs is also assigned as chairman of two NSC committees–the Committee on Foreign Intelligence (50 USC 402(h)(2)(D)) and the Committee on Transnational Threats (50 USC 402(i)(2)(E)). These assignments were made as part of the FY1997 Intelligence Authorization Act (P.L. 104-293); in signing it, President Clinton stated his concerns about the provisions relating to the establishment of the two NSC committees: "Such efforts to dictate the President's policy procedures unduly intrude upon Executive prerogatives and responsibilities. ("Statement on Signing the Intelligence Authorization Act for Fiscal Year 1997," October 11, 1996, *Weekly Compilation of Presidential Documents*, October 14, 1996, p. 2039). Other legislation placed the National Security Adviser on the President's Council on Counter-Narcotics (21 USC 1708(b)(1)(O)) and the Director of National Drug Control policy is required to work in conjunction with the Adviser "in any matter affecting national security interests." (21 USC 1703(b)(10))

[66] Shoemaker, *NSC Staff: Counseling the Council*, p. 115.

[67] Ivo H. Daalder and I.M. Destler, *In the Shadow of the Oval Office: Profiles of the National Security Adviser and the Presidents They Served—From JFK to George W. Bush* (New York: Simon & Schuster, 2009), pp. 318-319.

[68] See Steinberg, "Foreign Policy: Time to Regroup."

[69] For further background, see the PNSR website, http://www.pnsr.org. See also CRS Report RL34455, *Organizing the U.S. Government for National Security: Overview of the Interagency Reform Debates*, by Catherine Dale, Nina M. Serafino, and Pat Towell.

In: U.S. National Security: A New Strategy for... ISBN: 978-1-61761-436-1
Editors: Belinda M. White

Chapter 3

ORGANIZING FOR HOMELAND SECURITY: THE HOMELAND SECURITY COUNCIL RECONSIDERED*

Harold C. Relyea

SUMMARY

In the immediate aftermath of the terrorist attacks of September 11, 2001, President George W. Bush established the Office of Homeland Security and the Homeland Security Council (HSC). In his June 2002 proposal for a Department of Homeland Security, President Bush appeared to anticipate the continued operation of both of these entities. However, the Homeland Security Act of 2002, which mandated the new department, statutorily rechartered the HSC as an agency within the Executive Office of the President (EOP). Thereafter, the HSC disappeared from the public record, and its status today remains uncertain. Recently, some have called for the merger of the HSC with the National Security Council (NSC).

* This is an edited, reformatted and augmented edition of a United States Congressional Research Service publication, Report RS22840, dated November 26, 2008.

BACKGROUND

Among the initial actions taken by President George W. Bush in response to the September 11, 2001, terrorist attacks upon the World Trade Center and the Pentagon was the establishment of two new specialized agencies pursuant to E.O. 13228 of October 8, 2001.[1] These entities were the Office of Homeland Security (OHS), located within the Executive Office of the President (EOP) and headed by a new Assistant to the President for Homeland Security, and the Homeland Security Council (HSC), chaired by the President. Their creation demonstrated the commitment of the Administration to securing the homeland and combating terrorism. Each constituted a specialized coordination mechanism: the OHS for developing and coordinating the implementation of a comprehensive national strategy to secure the United States from terrorist threats or attacks; and the HSC for advising and assisting the President with respect to all aspects of homeland security. Each would be provided resources for carrying out its responsibilities. Moreover, the HSC would become the progenitor of, and repository for, new Homeland Security Presidential Directives (HSPDs).

Unlike the OHS, the HSC was not explicitly located within the EOP. Its basic membership was composed of the Vice President; the Secretaries of Defense, Health and Human Services, and Transportation; the Attorney General; the Director of Central Intelligence; the Directors of the Federal Bureau of Investigation and the Federal Emergency Management Agency; the Assistant to the President for Homeland Security; "and such other officers of the executive branch as the President may from time to time designate."[2] Prior to the October 8, 2001, issuance of E.O. 13228, President Bush, as he indicated in an October 5 letter to the Speaker of the House, authorized the transfer of $195.9 million from the emergency response fund, which had been established by statute for recovery from, and response to, the September 11 terrorist attacks,[3] $25.5 million of which was allocated for the establishment of OHS.[4] Some portion of these funds directly or indirectly also supported the HSC. The council held its initial meeting on October 29. That same day, the President inaugurated the issuance of HSPDs "that shall record and communicate presidential decisions about the homeland security policies of the United States." The initial directive concerned the organization and operation of the HSC; the second, also issued on October 29, pertained to combating terrorism through immigration policies;[5] and the third, issued March 11, 2002, established the Homeland Security Advisory System of five graduated threat conditions to apprise the executive branch and the nation of the risk of a terrorist attack.[6] Between October 2001 and January 2008, 23 HSPDs were issued, some in security classified status and some concurrently as National Security Presidential Directives, another presidential directives series.[7]

The Council Reconstituted

When the President proposed creating the Department of Homeland Security in June 2002, his plan indicated that he anticipated the continued operation of both the OHS and the HSC. According to the plan, "[t]he White House Office of Homeland Security and the Homeland Security Council will continue to play a key role, advising the President and coordinating a vastly simplified interagency process."[8] The President's plan, as introduced legislatively in the House on June 24, made no provision for a Homeland Security Council. As reported a month later and subsequently approved by the House, the legislation mandated a National Homeland Security Council chaired by the President and composed of the Vice President, eight specified Cabinet Secretaries, the Director of Central Intelligence, and such other officials as the President may designate. Joint meetings with the National Security Council were an option. This basic arrangement remained in the Homeland Security Act ultimately adopted by both houses of Congress and signed into law by the President on November 25, 2002.[9] Title IX of the statute reconstituted the HSC, located it within the EOP, and made it responsible for advising the President on homeland security matters; assessing the objectives, commitments, and risks of the United States in the interest of homeland security and making resulting recommendations to the President; and overseeing and reviewing homeland security policies of the federal government and making resulting recommendations to the President. Still chaired by the President, the council's membership was modified to include the Vice President, the Secretary of Homeland Security, the Attorney General, the Secretary of Defense, and such other individuals as the President may designate. Joint meetings with the National Security Council were an option.[10]

Thereafter, the HSC disappeared from the public record. It does not appear to have complied with requirements for *Federal Register* publication of such basic information as descriptions of its central organization; where, from whom, and how the public may obtain information about it; "statements of the general course and method by which its functions are channeled and determined"; and rules of procedure, substantive rules of general applicability, and statements of general policy.[11] No profile of, or descriptive information regarding, the HSC or its members and staff has appeared, to date, in the annual editions of the *United States Government Manual*.[12]

Similar conditions surround HSC resources. In late July 2003, House appropriators, in their report on the Departments of Transportation and Treasury and Independent Agencies Appropriations Bill, 2004, revealed that the Bush Administration had changed the "Office of Homeland Security" account, previously listed for the EOP, to an account for the "Homeland Security Council." The report also questioned the continued role of OHS, saying "it is not clear what work remains that cannot be effectively performed by the Department of Homeland Security." The account change also implied the shift of 66 staff personnel from OHS to the HSC, which the report questioned, "given the existence and support of the Department of Homeland Security." The committee cut the President's request of $8.3 million for the council to $4.1 million.[13] Senate appropriators declined to fund the HSC through the White House Office (WHO) account, as requested, and recommended the $8.3 million sought by the President for the council in a separate account for the HSC. They indicated that "the Homeland Security Council should be funded as a separate account, which is consistent with the budgetary treatment of its predecessor, the Office of Homeland Security."[14] The HSC, however, has no statutory authorization to receive appropriations. Conferees on the subsequent Consolidated Appropriations Act, 2004, when providing funds for agencies of the Executive Office of the President, allocated $7.2 million for the Homeland Security Council in the WHO account.[15] The House agreed to the conference report on December 8 and adjourned; the Senate completed action on the legislation on January 23, 2004, and the President signed the measure the following day.[16]

For FY2005, the President again requested funding for the HSC through the WHO account. House appropriators, in recommending almost $2.5 million for the council, did so in a separate account for the HSC, although the council has no authorization to receive appropriations. They were also unhappy about accounting for another HSC resource, saying,

> The Committee is disturbed that White House officials have failed to provide the Committee a definitive request for HSC staffing or budgetary resources for fiscal year 2005. Information providing [sic] for the hearing record states that the fiscal year 2005 budget includes "approximately" 40 full-time equivalent staffyears for direct HSC hires and 26 detailees, for a total of 66 staff. This estimate, although approximate, would be significantly

> above the level of onboard staff as of May 2004. In future years, the Committee expects the Executive Office of the President to be able to provide budget-quality estimates rather than approximations.[17]

Conferees on the subsequent Consolidated Appropriations Act, 2005, concurred on the amount approved by the House for the HSC, but provided the funding through the WHO account.[18]

When allocating funds for the WHO for FY2006, the report of the House Committee on Appropriations specified that "[t]he Salaries and Expenses account of the White House Office supports staff and administrative services necessary for the direct support of the President, including costs for the Homeland Security Council."[19] The Senate counterpart report, the related conference committee report, and the resulting Transportation, Treasury, Housing and Urban Development, the Judiciary, District of Columbia, and Independent Agencies Appropriation Act, 2006, did not contain such language and otherwise made no reference to the HSC. House appropriators reiterated their 2006 report language in their report on Transportation, Treasury appropriations for FY2007.[20] The counterpart Senate report did not contain such language. House appropriators again reiterated this language in their report on the Financial Services and General Government appropriations for FY2008.[21] Neither the counterpart Senate report nor the subsequent Consolidated Appropriations Act, 2008, contained such language or made any reference to the HSC.

In its July 2004 final report, the National Commission on Terrorist Attacks Upon the United States questioned the continued need for the HSC, saying, "To improve coordination at the White House, we believe the existing Homeland Security Council should soon be merged into a single National Security Council."[22] The executive director of the Project on National Security Reform, a non-partisan study group formed to analyze problems that inhibit interagency coordination, has also suggested such a merger.[23] In January 2008, at the time of presidential homeland security adviser Frances F. Townsend's departure from the White House, there was press speculation that her HSC staff would be transferred to the National Security Council (NSC) with her deputy supervising that staff in its new status.[24] There has been no public indication that such a staff shift has occurred. In mid-October 2008, it was reported that Ms. Townsend had twice attempted to eliminate her homeland security adviser position, initially in the spring of 2004 and again in early 2005, but had relented because the NSC staff was overwhelmed with dealing with warfare in Afghanistan and Iraq and could not assume oversight of domestic security matters.[25] In mid-November, speculation appeared in thepress that transition assistants for President-elect Barack Obama might recommend that the functions of the Homeland Security Council be transferred to another entity.[26]

The Council Reconsidered

The HSC was statutorily mandated in 2002 as an agency within the Executive Office of the President. It became the recipient of the staff of the OHS when that agency was eclipsed by the new Department of Homeland Security. It has, however, seemingly fallen short of achieving complete agency status due to its (1) seeming failure to publish such basic information as descriptions of its central organization; where, from whom, and how the public may obtain information about it; "statements of the general course and method by which its functions are channeled and determined"; and rules of procedure, substantive rules of general applicability, and statements of general policy;[27] (2) omission from the annual volumes of the *United States Government Manual*; and (3) lack of an authorization to receive appropriations as an agency within the EOP. Some have questioned the continued need for the HSC, viewing its functions as being more efficiently and effectively carried out by the National Security Council. Additionally, they maintain that such a merger might result in some economies, if only minimal, and might also provide an opportunity to eliminate what they see as the somewhat redundant HSPD series. The situation may be subject to congressional review, reconsideration, and action.

End Notes

[1] 3 C.F.R., 2001 Comp., pp. 796-802.
[2] Ibid., p. 800.
[3] 115 Stat. 220.
[4] The White House, letter to the Speaker of the House of Representatives, Oct. 5, 2001, Washington, DC (copy in the possession of the author).
[5] *Weekly Compilation of Presidential Documents*, vol. 37, Nov. 5, 2001, pp. 1568-1573.
[6] Ibid., vol. 38, Mar. 18, 2002, pp. 394-397.
[7] While unclassified HSPDs are usually available from the White House website upon issuance and are published in the *Weekly Compilation of Presidential Documents*, they are not published in the *Federal Register* or reproduced in the *Public Papers of the Presidents of the United States*. See CRS Report 98-611, *Presidential Directives: Background and Overview*, by Harold C. Relyea.

[8] U.S. White House Office, *The Department of Homeland Security* (Washington: June 2002), p.3.
[9] 116 Stat. 2135.
[10] 116 Stat. 2258.
[11] 5 U.S.C. §552(a).
[12] A profile of the council is available from the White House website at [http://www.whitehouse.gov/hsc/].
[13] U.S. Congress, House Committee on Appropriations, *Departments of Transportation and Treasury and Independent Agencies Appropriations Bill, 2004*, a report to accompany H.R. 2989, 108th Cong., 1st sess., H.Rept. 108-243 (Washington: GPO, 2003), p. 163.
[14] U.S. Congress, Senate Committee on Appropriations, *Transportation, Treasury and General Government Appropriations Bill, 2004*, a report to accompany S. 1589, 108th Cong., 1st sess., S.Rept. 108-146 (Washington: GPO, 2003), p. 135.
[15] See U.S. Congress, House Committee of Conference, *Making Appropriations for Agriculture, Rural Development, Food and Drug Administration, and Related Agencies for the Fiscal Year Ending September 30, 2004, and for Other Purposes*, conference report to accompany H.R. 2673, 108th Cong., 1st sess., H.Rept. 108-401 (Washington: GPO, 2003), p. 1015; *Congressional Record*, daily edition, vol. 149, Nov. 25, 2003, p. H12409; 118 Stat. 321.
[16] 118 Stat. 3.
[17] U.S. Congress, House Committee on Appropriations, *Departments of Transportation and Treasury and Independent Agencies Appropriations Bill, 2005*, report to accompany H.R. 5025, 108th Cong., 2nd sess., H.Rept. 108-671 (Washington: GPO, 2004), p. 129.
[18] 118 Stat. 3246.
[19] U.S. Congress, House Committee on Appropriations, *Departments of Transportation, Treasury, and Housing and Urban Development, the Judiciary, District of Columbia, and Independent Agencies Appropriation Bill, 2006*, report to accompany H.R. 3058, 109th Cong., 1st sess., H.Rept. 109-153 (Washington: GPO, 2005), p. 133.
[20] U.S. Congress, House Committee on Appropriations, *Departments of Transportation, Treasury, and Housing and Urban Development, the Judiciary, District of Columbia, and Independent Agencies Appropriation Bill, 2007*, report to accompany H.R. 5576, 109th Cong., 2nd sess., H.Rept. 109-495 (Washington: GPO, 2006), p. 178.
[21] U.S. Congress, House Committee on Appropriations, *Financial Services and General Government Appropriations Bill, 2008*, report to accompany H.R. 2829, 110th Cong., 1st sess., H.Rept. 110-207 (Washington: GPO, 2007), p. 33.
[22] U.S. National Commission on Terrorist Attacks Upon the United States, *The 9/11 Commission Report* (Washington: GPO, 2004), p. 406.
[23] James R. Locher III, executive director, Project on National Security, briefing for CRS staff, Mar. 6, 2008, Washington, DC.
[24] "Fond Farewell," *Washington Times*, Jan. 4, 2008, p. A11.
[25] Chris Strohm, "Ex-Bush Adviser Cites Need for New Security Apparatus," *CongressDaily PM*, Oct. 15, 2008, available at [http://www.nationaljournal.com/congressdaily/print/_friendly.php?ID=cdp_20081015_8843].
[26] Karen DeYoung, "Naming National Security Team Will Be a Priority for Obama," *Washington Post*, Nov. 19, 2008, p. A8.
[27] 5 U.S.C. §552(a).

INDEX

D

N

O

P

R

S